GNSS Receivers for Weak Signals

The GNSS Technology and Applications Series

Elliott Kaplan and Christopher Hegarty, Series Editors

Understanding GPS: Principles and Applications, Second Edition, Elliott Kaplan and Christopher Hegarty, editors

Introduction to GPS: The Global Positioning System, Second Edition, Ahmed El-Rabbany

GNSS Receivers for Weak Signals, Nesreen I. Ziedan

Applied Satellite Navigation Using GPS, GALILEO, and Augmentation Systems, Ramjee Prasad and Marina Ruggieri

Geographical Information Systems Demystified, Stephen R. Galati

Digital Terrain Modeling: Acquisition, Manipulation, and Applications, Naser El-Sheimy, Caterina Valeo, and Ayman Habib

GNSS Receivers for Weak Signals

Nesreen I. Ziedan

ARTECH
HOUSE

BOSTON | LONDON
artechhouse.com

A catalog record for this book is available from the U.S. Library of Congress.

A catalogue record for this book is available from the British Library.

ISBN 10: 1-59693-052-7
ISBN 13: 978-1-59693-052-0

Cover design by Yekaterina Ratner

Published by ARTECH HOUSE, Inc.
685 Canton Street
Norwood, MA 02062

10 9 8 7 6 5 4 3 2 1

Contents

Preface

Many applications have required the positioning accuracy of a Global Navigation Satellite System (GNSS). Some applications exist in environments that attenuate GNSS signals, and, consequently, the received GNSS signals become very weak. Examples of such applications are wireless device positioning, positioning in sensor networks that detect natural disasters, and orbit determination of geostationary and high earth orbit (HEO) satellites. Conventional GNSS receivers are not designed to work with weak signals. This book presents novel GNSS receiver algorithms that are designed to work with very weak signals.

Wireless devices receive signals with very low power, typically below a carrier-to-noise ratio (C/N_0) of 25 dB-Hz, due to attenuation incurred from the surrounding environments, such as operating indoors or under heavy vegetation. In addition, such devices usually have small processors, small memory size, and limited battery life.

Geostationary and HEO satellites, which are located above the GNSS constellation, receive GNSS signals with very low power due to the high path loss of these signals. This is because the GNSS antenna is oriented toward the earth, so the transmitted signals are not accessible to any receiver above the transmitting GNSS satellite. However, the antenna pattern enables those signals to span the space beyond the earth, and so the signals become reachable by receivers on the opposite side of the earth. The number of visible satellites is small, however, and each satellite is available for a short period.

This book focuses on the development of efficient GNSS receiver algorithms to work with weak signals under various dynamic conditions. The algorithms do not require any assisting information from outside sources (e.g., networks); thus, they can be implemented in stand-alone GNSS receivers. The developed algorithms address all of the receiver main functions, which include signal acquisition, fine acquisition, bit synchronization, data detection, code and carrier tracking, and navigation message decoding.

Fifteen different algorithms have been developed. These algorithms are as follows: two algorithms for the acquisition of weak signals; one algorithm for the acquisition of weak signals in the presence of strong interfering signals; one algorithm for high dynamic acquisition; fine acquisition and high dynamic fine acquisition algorithms; a bit synchronization and data-detection algorithm; code- and carrier-tracking algorithms that work under low- and high-dynamic conditions; one algorithm to detect and correct large carrier-tracking errors; one algorithm to detect and correct large code-tracking errors; one algorithm to deal with large sudden changes in the receiver dynamics; preamble identification, subframe identification, and navigation message decoding algorithms.

The processing and memory requirements have been considered as main criteria in the design of the algorithms to enable them to fit the limited resources of some of the targeted applications.

Overview of the Book's Organization

This book is organized as follows:

- Chapter 1 provides some introductory material about the GNSS, the Global Positioning System (GPS) signal structure and processing, different acquisition and tracking techniques, an overview of the new GPS signals and the Galileo navigation system, a summary of some weak-signal applications, and some technical concepts that have been applied to the algorithms introduced in this book.
- Chapter 2 presents a detailed derivation of the effect of the Doppler shift on code length. It also provides models for the received and locally generated signals. These models are used to derive the models used in the various stages of the GPS receiver signal processing. In addition, a clock noise model is provided.
- Chapter 3 presents the developed weak-signal acquisition algorithms. The algorithms also address the problems of acquiring weak signals in the presence of strong interfering signals and high dynamics acquisition. The analysis and derivation of the probabilities of false alarm and detection and the calculation of the acquisition threshold are presented.
- Chapter 4 presents the fine acquisition and bit synchronization and data-detection algorithms. These algorithms are based on the Viterbi algorithm (VA) and the extended Kalman filter (EKF). Fine acquisition in the case of a high Doppler rate error is also presented.

- Chapter 5 presents the code- and carrier-tracking algorithms. These algorithms are based on EKF approaches. Several EKF designs are introduced, including first-order, second-order, and square root EKFs. Also, some methods are introduced to increase the time to lose lock. In addition, a navigation message decoding algorithm is presented; this algorithm utilizes the navigation message structure to enable its decoding for signals with high bit error rate (BER).
- Chapter 6 presents an overall summary of the algorithms and their functionalities. It also provides an analysis and performance comparison between the algorithms presented in this book and some conventional aided and unaided acquisition and tracking algorithms.

1

Overview of GNSS Principles and Weak-Signal Processing and Techniques

1.1 Introduction

There are currently two operational Global Navigation Satellite Systems (GNSS) [1–3]: the Global Positioning System (GPS), which is operated by the U.S. government; and the Global Orbiting Navigation Satellite System (GLONASS), which is operated by the government of the Russian Federation. New GPS signals have been developed and will start transmission with the modernized Blocks IIR-M and IIF satellites. In addition, a new GNSS, Galileo, is being developed and implemented in a collaborative effort between the European Union (EU) and the European Space Agency (ESA); it will be under civilian control.

Many new GNSS applications have been emerging with operational requirements beyond the capability offered by conventional GNSS receivers. This has motivated advances in the GNSS receiver algorithms to fit different applications requirements.

This chapter provides some introductory material about GPS, different acquisition and tracking techniques, and some technical concepts that have been applied to the algorithms introduced in this book. This will give the reader a general overview of the subjects relevant to this book.

In this book, the terms *algorithm* and *module* refer to two different things. An algorithm is a method to do something. A module is an object or entity; it refers to an ensemble of algorithms that interact to perform a task.

This chapter is organized as follows. Section 1.2 discusses the GPS components. Section 1.3 discusses the principles of user position and velocity determination. Section 1.4 gives an overview of the GPS signal structure. Section 1.5 gives an overview of the GPS navigation message structure. Section 1.6 gives an overview of the signal processing done prior to the correlation. Section 1.7 summarizes the techniques generally used for signal acquisition. Section 1.8 gives an overview of synchronization techniques and summarizes some synchronization approaches used with GPS. Section 1.9 gives an overview of the conventional GPS tracking loops, and then it summarizes GPS tracking techniques. Section 1.10 summarizes the encoding and decoding of the GPS navigation message. Section 1.11 gives a general overview of the modernized GPS and the new Galileo navigation system. Section 1.12 presents a summary of some weak-signal applications. Section 1.13 presents an overview of some technical concepts used in the development of the algorithms introduced in this book. Section 1.14 discusses some problems associated with using weak signals for positioning, and then it summarizes the algorithms introduced in this book.

1.2 GPS Components

1.2.1 Space Segment

The GPS satellite constellation consists of 24 primary satellites [1, 2] distributed in six orbital planes inclined by 55° with respect to the equator. Each orbit has four primary satellites and a spare slot for one additional satellite. The satellite constellation has a period of 12-hour sidereal time (11 hours 58 minutes). Each satellite is identified by a letter and a number; the letter refers to the satellite orbital plane (A–F), and the number refers to the satellite order in the plane (1–5).

A three-dimensional position determination requires at least four visible satellites. However, if the altitude is known, then a two-dimensional position determination can be achieved using at least three visible satellites. The primary 24 satellites cover the earth such that at least four satellites are visible at any time and at any point on the earth. There can, however, be up to twelve visible satellites.

The first model of the GPS was launched between 1978 and 1985 and identified as Block I. Following that, there have been several new models identified as Blocks II, IIA, and IIR. A modernized GPS signal structure will be transmitted by two new models identified as Blocks IIR-M and IIF.

1.2.2 Control Segment

The control segment consists of three main parts: the master control station, monitor stations, and ground antenna upload stations. Each monitor station has several GPS receivers, with cesium clocks, that continuously track the visible GPS satellites. The measurements from each monitor station are time tagged and transmitted to the master control station. The master control station processes the received measurements to estimate the navigation data parameters, such as the satellite orbits, clock errors, and satellite health, for each satellite. The master control station then computes and updates the navigation message for each satellite and transmits it to the ground antenna stations, which upload the navigation message to the GPS satellites. The navigation message is then transmitted with each satellite signal, which in turn is received by the GPS receivers. Both the monitor stations and the ground antenna upload stations are operated remotely from the master control station.

1.2.3 User Segment

The user segment defines the GPS receiver equipments. There have been many new applications of GPS positioning, such as wireless devices positioning, car navigation, earthquake monitoring, orbit determination of geostationary and high earth orbit (HEO) satellites, and tracking during ionospheric scintillation. Some applications require that the GPS receiver be able to operate in environments that attenuate the GPS signal power to very weak levels. These new applications have motivated advances in GPS receivers. There are many types of hardware and software receivers, with different capabilities and functionalities, that can fit different applications requirements.

In general, the GPS receiver's main functions include down conversion and sampling of the received radio frequency (RF) signal, GPS visible satellite search and acquisition, code and carrier tracking of the acquired signals, decoding of the navigation message, and calculation of the navigation solution.

1.3 Principles of User Position, Velocity, and Time Determination

Position determination using a GNSS is based on measuring the distances between an observer and several satellites at known locations, then solving for the observer's position. The measured distance is defined as the pseudorange, ρ. Define the known position vector of a satellite i as $r_i = [x_i \;\; y_i \;\; z_i]^T$ and the unknown position vector of an observer as $r_u = [x_u \;\; y_u \;\; z_u]^T$. By measuring pseudorange ρ from three different satellites, the user position can be determined by solving

$$\rho_i = |r_i - r_u| = \sqrt{(x_i - x_u)^2 + (y_i - y_u)^2 + (z_i - z_u)^2} \tag{1.1}$$

The pseudorange is calculated from the signal's time of travel from a satellite to the observer. Since radio signals travel at the speed of light, $c = 299{,}792{,}458$ m/s, ρ is calculated from

$$\rho = \text{ time of travel} \times \text{speed of light} \tag{1.2}$$

A signal's time of travel is calculated as the difference between the transmit time from a satellite and the receive time at the receiver. Thus, to obtain an accurate time-of-travel measurement, the clocks of the satellite and the receiver must be synchronized. However, because of the high cost of accurate clocks and the need to have a low-cost GPS receiver, the receiver clock is usually not accurate. The receiver clock bias is treated as an unknown, τ_{bias}. In addition, there are other sources of errors, like the receiver measurement noise and atmospheric delays. Define those errors with one term, n_ρ. Thus, (1.1) is modified to be

$$\rho_i = |r_i - r_u| + c\,\tau_{bias} + n_{\rho_i} \tag{1.3}$$

If n_ρ is assumed to be negligible, then at least four satellites are needed to solve for the user position. This problem can be solved using an extended Kalman filter (EKF). The visible satellites' geometry affects the quality of the position estimate. The relationship between the pseudorange accuracy and the user position accuracy is a function of the geometry of the visible satellites used to solve (1.3). The quality of the satellites' geometry is expressed by *dilution of precision* (DOP). A comprehensive discussion of the satellites' geometry and DOP can be found in [1, Chs. 5, 11] and [2, Ch. 5].

The relative motion between a satellite and observer results in a change in the observed carrier frequency. The amount of change in the carrier frequency defines the Doppler shift, f_d. The Doppler shift can be expressed as [1]

$$f_{d_i} = -\left(\frac{v_i - v_u}{c} \cdot \mathbf{1}_i\right) f_L \tag{1.4}$$

where $\mathbf{1}_i = (r_i - r_u)/(|r_i - r_u|)$ is the line of sight between the satellite and the observer; v_i and v_u are the satellite and the observer velocity vectors, respectively; and f_L is the carrier frequency. The Doppler shift is related to the change in the pseudorange, or the pseudorange rate $\dot{\rho}$, by

$$f_{d_i} = -\frac{\dot{\rho}_i}{\lambda} \tag{1.5}$$

where $\lambda = c/f_L$ is the wavelength. By taking into consideration the clock bias rate of the receiver and the measurement error, the pseudorange rate is obtained from

$$\dot{\rho}_i = (v_i - v_u) \cdot \frac{r_i - r_u}{|r_i - r_u|} + c\,\dot{\tau}_{bias} + n_{\dot{\rho}_i} \tag{1.6}$$

The Doppler shift can be obtained from the output of the tracking module. The satellite velocity can be obtained from the navigation message. The line of sight can be obtained from the estimated user position. The user velocity can be obtained by solving (1.6), using approaches such as EKF, least-squares, or weighted least-squares.

1.4 GPS Signal Structure

The GPS satellites of Block I through Block IIR transmit two types of GPS signals on the L1 and L2 carrier frequencies. These signals are a civil signal on the L1 carrier frequency, known as the coarse/acquisition (C/A) signal, and a military signal on the L1 and L2 carrier frequencies, known as the precession (P[Y]) signal. The Block IIR-M is designed to transmit an additional civil signal on the L2 carrier frequency. The block IIF is designed to transmit another two additional civil signals on the L5 carrier frequency. The L1, L2, and L5 carrier frequencies belong to the ultra high frequency (UHF) band. Those frequencies have values of 1,575.42 MHz, 1,227.6 MHz, and 1,176.45 MHz, respectively. Each GPS satellite has atomic standard clocks of 10.23 MHz. Each of the GPS carrier frequencies is generated as an integer multiple of the 10.23-MHz satellite reference clock. The integer multiples for the L1, L2, and L5 carrier frequencies are 154, 120, and 115, respectively.

Each signal consists of three components: an RF carrier frequency, a pseudorandom noise (PRN) code that serves as a ranging code, and a navigation message that contains the ephemeris and the almanac data needed for the calculation of the navigation solution. The repetitions of the carrier frequencies, the codes, and the data message are all related.

Each satellite transmits a unique PRN code with each signal type. The PRN codes are used to allow the receiver to identify each satellite and, consequently, to calculate the time of travel of the signal from the satellite to the receiver. The PRN codes used with each signal type are a subset of a family of Gold codes. The Gold codes are product codes of two maximal length codes. The PRN codes belong to a subset called *preferred pair Gold codes.* The PRN codes for each signal type are chosen to give good cross-correlation properties. The C/A code has a length of 1,023 chips and a transmission rate of 1.023 Mchip/s (i.e., it repeats every 1 ms). The P code has a length of 6.1871×10^{12} chips and a transmission rate of 10.23 Mchip/s (i.e., it repeats every 7 days).

The navigation message on the L1 carrier frequency has a transmission rate of 50 Hz. The PRN code and the navigation message are combined together, and they modulate the carrier frequency using a binary phase shift keying (BPSK) scheme. The start of each data bit is synchronized with the start of the C/A code. Each data bit has exactly 20 C/A code periods.

The GPS L1 signal has the C/A signal in phase and the P signal in quadrature phase. It has the form

$$S(t) = \sqrt{2P_{C/A}}\, C_{C/A}(t)\, d(t)\, \cos(\phi + 2\pi\, f_{L1}\, t) \\ + \sqrt{2P_P}\, C_P(t)\, d(t)\, \sin(\phi + 2\pi\, f_{L1}\, t) \qquad (1.7)$$

where $P_{C/A}$ and P_P are the C/A and P signal powers, respectively; $C_{C/A}$ and C_P are the C/A and P codes, respectively; d is the navigation data; f_{L1} is the L1 carrier frequency; and ϕ is an initial phase. The C/A signal is 3 dB stronger than the P signal.

1.5 GPS Navigation Message Structure

The GPS navigation message transmitted by each satellite contains the ephemeris of that satellite and the almanac data of all the satellites in the GPS constellation. The ephemeris includes such data as the satellite's position, velocity, and clock bias parameters. The almanac includes reduced-precision data about all of the satellites in the GPS constellation.

The navigation message transmitted on the L1 carrier frequency is 25 consecutive 1,500-bit-long frames. Each frame is made up of five subframes [4]. Figure 1.1 shows the structure of a subframe. Each subframe has a duration of 6 seconds, and each consists of 10 words. Each word consists of 30 bits. Subframes 1 to 3 contain the ephemeris and repeat in every frame (i.e., they repeat every 30 sec.), while subframes 4 and 5 contain the almanac and repeat every 25 frames (i.e., they repeat every 12.5 minutes). Each word, in each subframe, contains a 6-bit parity.

The first two words of each subframe are the telemetry (TLM) and handover (HOW) words. The TLM word begins with an 8-bit preamble, followed by a

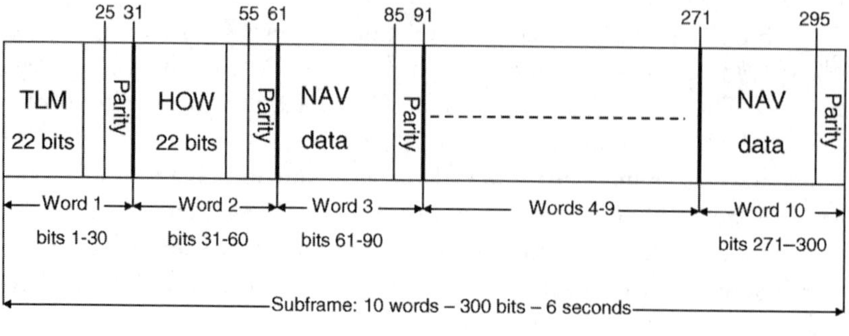

Figure 1.1 Structure of a subframe of the navigation message.

14-bit TLM message, two reserved bits, and then a 6-bit parity. The HOW word begins with the 17 most significant bits of the 19-bit time-of-week (TOW) count [4]. Bits 20 to 22 of the HOW word provides the ID of the subframe of which this word is a part. The ID codes for subframes 1 to 5 are 001, 010, 011, 100, and 101.

The ephemeris parameters are contained in words 3 to 10 of subframes 1 to 3, while the almanac parameters are contained in words 3 to 10 of subframes 4 and 5.

1.6 Precorrelation Signal Processing

1.6.1 Frequency Down Conversion

The GPS signal goes through several stages before it is processed to acquire and track the visible satellites. The RF GPS signal is received by a circularly polarized antenna, which can receive signals from any direction. The signal then goes through the receiver front end, which contains a low-noise amplifier (LNA) and some filters to reject multipath and interfering signals. The front end elements set the noise figure of the receiver and, consequently, the carrier-to-noise ratio (C/N_0) of the received signal.

Following from (1.7), the received RF L1 C/A signal for one satellite has the form

$$r_{RF}(t) = \sqrt{2P_{C/A}}\, C_{C/A}(t)\, d(t)\ \cos(\phi_0 + 2\pi\,(f_{L1} + f_d)\, t) + n_{RF}(t) \qquad (1.8)$$

where ϕ_0 is the initial received phase; f_d is the Doppler shift; and $n_{RF}(t)$ is a Gaussian noise with zero mean and σ_n^2 variance. The received RF signal is converted into an intermediate frequency (IF) signal by mixing the RF signal with a local oscillator. The local oscillator has the form

$$LO(t) = 2\cos(2\pi\, f_{LO}\, t) \qquad (1.9)$$

where f_{LO} is the frequency of the local oscillator. The mixing process will result in a signal with some harmonics plus upper and lower sidebands [i.e., sidebands with center frequencies of $(f_{L1} + f_{LO})$ and $(f_{L1} - f_{LO})$, respectively]. A bandpass filter is used to remove the harmonics, and a lowpass filter is used to remove the upper sideband. The center frequency of the lower sideband is the IF carrier frequency, f_{IF}. The resulting signal will have the form

$$r_{IF}(t) = \sqrt{2P_{C/A}}\, C_{C/A}(t)\, d(t)\ \cos(\phi_0 + 2\pi\,(f_{IF} + f_d)\, t) + n_{IF}(t) \qquad (1.10)$$

where $f_{IF} = f_{L1} - f_{LO}$.

1.6.2 Sampling and Quantization

The received signal in (1.10) is an analog signal that can take any numerical value. This signal is converted into digital form by sampling it at discrete times, using automatic gain control (AGC) to maintain the signal levels within an acceptable range and quantizing the signal to take discrete values.

Two different sampling techniques have been used with GPS signals: baseband sampling and IF sampling [1, 2]. IF, or bandpass sampling, combines the sampling and the down-conversion functions. In order to reconstruct a signal fully from the discrete samples, the sampling frequency, f_s, has to satisfy the Nyquist theorem and prevent aliasing. In baseband sampling, f_s has to be greater than twice the highest frequency component of the signal. In IF sampling, f_s has to be greater than twice the bandwidth of the IF signal.

The quantization process maps the signal levels to the nearest quantization threshold values. The number of the thresholds, N_{γ_Q}, is determined from the number of discrete signal levels used to represent the received signal. If A_{max} and A_{min} define the highest and the lowest values of the signal at the input of the quantizer, then the resolution of the quantization process is $(A_{max} - A_{min})/(N_{\gamma_Q} - 1)$.

The sampled received signal is expressed as

$$r_k = A\, C_{C/A_k}\, d_k \cos(\phi_0 + 2\pi\,(f_{IF} + f_d)\, t_k) + n_k \tag{1.11}$$

where t_k is the sampling time, and A is the signal amplitude. The signal in (1.11) is correlated with a replica code, then used within different modules of the receiver to acquire and track a satellite. Different signal models used at different stages in a GPS receiver are discussed in detail in Chapter 2.

1.7 Overview of Acquisition Techniques

GPS signal acquisition starts with a search process. The goal of the acquisition is to find the visible satellites; the C/A code delay, τ; and the Doppler shift, f_d. The receiver down-converts the RF signal into an IF signal. The IF signal is digitized, sampled, and then passed to the acquisition module.

The code delay and Doppler shift of each satellite signal are unknown; thus, the acquisition module performs two-dimensional search on all possible C/A code delays, N_τ, and all possible Doppler shifts, N_{f_d}. The acquisition module generates a replica C/A code of a satellite and correlates it with the IF received signal at all possible code delays and Doppler shifts. Then, it integrates the result coherently over T_I ms and incoherently over a number L of the T_I coherent results; the length of T_I and the number L depend on the received signal strength. If the relative code delay between the local code and the received code is less than the duration of one code chip, T_{chip}, and if the searched Doppler bin is located less $1/T_I$ from the received Doppler shift, then the received IF signal and the locally generated one will correlate with a value proportional to the errors in the code delay and the Doppler shift; the correlation reaches a maximum value at zero errors. Otherwise, the local and the received signals will not correlate. The acquisition is concluded if the power of the total integration crosses a predefined acquisition threshold, γ. The code delay and the Doppler shift estimates will be those that generated the peak power.

The conventional hardware acquisition approach [1, 2] searches for a satellite sequentially at each possible code delay and Doppler shift. At each step, it generates a Doppler-compensated replica code shifted by a possible code delay, multiplies it by the received signal, and integrates the result. The search continues until the power of an integration output exceeds γ.

Circular correlation [5, 6] is an alternate approach that is based on fast Fourier transform (FFT) methods. It calculates the coherent integration for all the code delays, at each Doppler shift, in the same processing step. Thus, it requires less processing as compared to the conventional approach. First, the FFT is calculated separately for the received signal and the Doppler-compensated locally generated one. Then, the coherent integration is found as the inverse FFT (IFFT) of the received FFT signal multiplied by the complex conjugate of the local FFT signal. The code delay and Doppler shift that generate the largest power, given that this power exceeds γ, will be those closest to the received values. In [7], the processing required for the circular correlation approach is reduced by using a single sideband approach in which the IFFT is found for only the first half of the points. This works because most of the power of the spectrum resulting from multiplying the local and the received FFT exists in those points. This, however, reduces the code-delay resolution by half. The paper suggests regaining the lost resolution by applying a three-point curve fitting using the peak signal point and its two neighboring points.

The delay-and-multiply approach [5, 7] eliminates all of the Doppler bins. The proposed approach is a modified version, suitable for real received signals, of an approach developed in [8], which is suitable for complex code division multiple access (CDMA) signals. The idea behind that approach is that if the complex received signal is multiplied by the complex conjugate of a delayed version of itself, then this will generate a signal that is independent of the Doppler shift. Moreover, the code of the new signal belongs to the same family of Gold codes as the received code, and it has the same delay as the received code. If $s(t)$ is the received signal with a spreading code $C(t)$, and $s(t-\tau_i)$ and $C(t-\tau_i)$ are delayed versions by an arbitrary value τ_i, then

$$s(t)s(t-\tau_i)^* = C(t)\, C(t-\tau_i)^*\, e^{j2\pi f\tau_i} \tag{1.12}$$

where $*$ defines the complex conjugate of the signal. The new code is $C_n = C(t)\, C(t-\tau_i)^*$. The beginning point of the code can be found by correlating a local version of the new code with the signal resulting from (1.12). The Doppler shift can be found by applying an FFT to the correlation result. This approach, however, has low sensitivity due to the multiplication process, so it is only suitable for signals with high signal-to-noise ratio (SNR). The modified approach follows similar steps, except because the received GPS signal is a real signal and not a complex one, there is a restriction on the amount of delay τ_i; τ_i is chosen to make $\cos(2\pi f \tau_i)$ closest to 1.

Weak-signal acquisition requires long coherent and incoherent integration. Increasing the coherent integration improves sensitivity, but it causes an increase in the number of searched Doppler bins and is limited because of the unknown data bits and bit edges. The incoherent integration averages the absolute values, so it does not suffer from the same limitations as the coherent integration, but it is less sensitive.

In [9], an approach is introduced to reduce the processing time when using long T_I. The IF received signal is multiplied by an IF Doppler-compensated local signal. The output is divided into a number of groups, each spanning 1 ms of length. The groups are added together (i.e., points at the same index in each group are added together). Then, a circular correlation is performed on the addition result. The circular correlation will be done over N_τ points instead of N_τ T_I points.

An approach that processes long data coherently with fewer operations, double block zero padding (DBZP), is discussed in [5, 9, 10]. A modified version of this approach is developed in one of the acquisition algorithms introduced in Section 3.3. The modified version circumvents some problems in the original DBZP approach that arise when very long coherent and incoherent integrations are used. Figure 1.2 illustrates the DBZP processing. DBZP first converts the received IF signal into baseband. Using a predetection integration time (PIT) of T_I ms, the T_I-long baseband samples are arranged into blocks. The number of blocks is equal to the number of Doppler bins, N_{f_d}. The Doppler bin separation is equal to $1/T_I$. Define the block size by S_{block}. Each two adjacent blocks of the received samples are combined into one block. The last block is combined with additional samples of size S_{block} that are taken beyond the T_I ms' worth of samples. This results in the same number of blocks as the original number but with double the size. T_I-long samples of the replica C/A code are arranged into the same number of blocks. Each block of the replica C/A code is padded with S_{block} zeros. Circular correlation is calculated using each two corresponding blocks of the received and the local signals. Only the first S_{block} points of the circular correlation result are preserved from each block. The preserved points, in each block, represent a partial coherent integration of length T_I/N_{f_d} ms at S_{block} possible code delays. The preserved points are arranged into a matrix M_c of initial size equal to $N_{f_d} \times S_{block}$; each column contains the points located at the same index in each block. To cover all possible N_τ code delays, the process

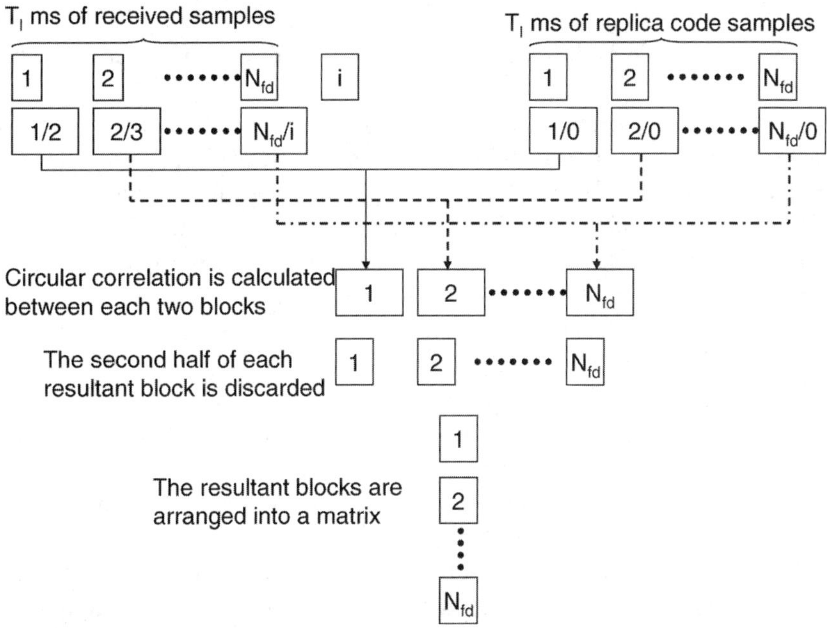

Figure 1.2 Illustration of DBZP processing.

is repeated. In each repetition, the blocks of the replica code are shifted one step such that the first block becomes the last one, while the second becomes the first, and so on. Following each repetition, the result is appended to M_c. At the end, M_c will have a size of $N_{f_d} \times N_\tau$. Then, the FFT is calculated for each column of M_c; this will be an N_{f_d} FFT-points operation. The result represents a T_I-ms coherent integration. Each row represents the coherent integration at a possible Doppler shift, and each column represents the coherent integration at a possible code delay. The FFT component that has the maximum value and crosses the acquisition threshold γ will correspond to the estimated code delay and Doppler shift. The advantage of this approach is that it calculates the coherent integration for all possible Doppler shifts in the same operation. Therefore, it requires much less processing compared to the original circular correlation approach, which calculates the coherent integration at each possible Doppler shift separately using FFT/IFFT.

Some alternatives are also presented in [9]. One approach is similar to DBZP, but instead of combining each two adjacent blocks into one block with double the size, the corresponding points of each two adjacent blocks are added together. This approach requires less processing time but has more loss than the original. Another approach converts the received IF signal into baseband and then squares it. The result is a continuous wave with double the Doppler shift. The result is divided into groups, each of which is averaged into one point to reduce the noise. Then, FFT is performed on the averaged points to find the Doppler shift. Once the Doppler shift is found, the C/A code delay can be found using an FFT process.

In [11], some analysis is provided for the case when T_I is 40 ms, the data bit values are not known, and there is a navigation data transition. In general, a data transition causes multiple peaks around the correct Doppler bin. This observation is utilized either by adding the amplitudes of the adjacent spectral lines to achieve an increase in SNR of 1.5 dB or by adding the values of the adjacent spectral lines coherently to achieve an increase in the SNR of about 3 dB.

In [12], a method is introduced to reduce the processing time by reducing the number of points over which the circular correlation is calculated. The number of points is reduced by adding the samples that belong to the same chip. Since the start of the chips is not known, however, five different averages are calculated. Each average starts at a possible first sample of a chip.

In [13], an acquisition method is introduced that uses T_I of 10 ms, without knowing the values of data bits or the positions of the bit edges. The method calculates two independent accumulations. The start of each accumulation is separated from the other one by 10 ms. It divides each 20-ms sample into two groups of 10 ms each. It performs coherent integration separately for each 10 ms. Then, it adds the result of each corresponding 10 ms incoherently. This method guarantees that at least one group will not have a data transition, so its acquisition result will not be degraded compared to that of the other group.

In [14], the problem of acquiring weak signals with a GPS software receiver is addressed. Two techniques, which can be used to increase the T_I and remove the data transition effect, are introduced. The first technique is similar to the one in [13]. The second one uses 20-ms coherent integration, followed by incoherent integration. Since the data transition is not known, the method forms 20 groups of accumulations, each starting at a possible bit edge position. The group that contains the maximum power in one of its cells will correspond to the correct bit edge position. The code delay and Doppler shift estimates are determined from the cell that generated that maximum. The algorithm was able to acquire signal with 21 dB-Hz using 199 incoherent integrations (i.e., a total of 199 data bits). Also discused are some methods that can be used to decrease the processing associated with the increase in the number of Doppler bins.

In [15], an analysis of the coherent integration capability to acquire weak signals is provided. The analysis assumes that the effect of the navigation data is removed and that rough Doppler shift information is available from an assisting source. The results show that at least 8-ms coherent integration is needed to acquire a 32 dB-Hz signal; at least 200-ms coherent integration is needed to acquire a 22 dB-Hz signal; at least 400-ms coherent integration is needed to acquire a 17 dB-Hz signal; and at least 800-ms coherent integration is needed to acquire a 12 dB-Hz signal.

In [16], some simulation results of the coherent integration capability, taking into consideration the effect of the rate of change of the Doppler shift, are provided. The results indicate that very long coherent integration cannot be used to detect very weak signals unless the rate of change of the Doppler shift is compensated and that it is better to use the incoherent integration in this case.

1.8 Overview of Bit Synchronization and Data-Estimation Techniques

Timing recovery, or bit synchronization, techniques are divided into three main categories [17, 18]:

1. *Data aided (DA):* A known sequence of symbols is transmitted along with the data; this sequence is utilized for bit synchronization.
2. *Decision directed (DD):* The data are estimated prior to the bit synchronization, and then they are used in the synchronization module. DD is similar to DA, except that there might be errors in the estimated data.
3. *Non–data aided (NDA):* No data are utilized for the bit synchronization; estimation techniques based on maximum likelihood (ML) approaches are usually used.

Bit synchronization can be performed using (1) a feedback scheme, where an error signal is generated to control the output of a voltage control oscillator (VCO), or (2) a feedforward scheme, where the timing information is directly extracted from the received data, and the transmitted symbols are constructed by interpolating the signal samples.

An example of a feedback scheme is the data transition tracking loop (DTTL) [18]. This loop generates an error signal proportional to the synchronization error between the estimated and the true bit boundary. The error signal is calculated from the output of in-phase and quad-phase integrate and dump channels, which have the received signal as their input. The error signal is then used to adjust the instantaneous frequency that controls a timing pulse generator, which, in turn, controls the start of the integration time in both channels. The in-phase channel integrates the input signal between the start and the end of an estimated bit boundary. The quad-phase channel performs the integration over a window with its middle point located at the estimated bit boundary. The value of the error signal is taken as the output of the quad-phase channel; the sign of the error signal is taken as the sign resulting from multiplying the output of both channels together. The sign of the error signal indicates whether the estimated bit boundary leads or lags behind the true one. In [19, 20], a modified DTTL is presented; it has enhanced performance at low SNR. Examples of feedforward schemes are presented in [21, 22].

Bit synchronization algorithms for GPS include the histogram method [1] and ML type estimation [23, 24]. The histogram method uses a counter at each possible bit edge position. It counts the sign changes between each two 1-ms consecutive correlated signals, at intervals of one data bit, and then it compares the result to two thresholds. All the counters are reinitialized if two of them cross the lower threshold. The bit edge position is concluded if a counter reaches the upper threshold. The performance of this algorithm depends on the carrier-to-noise ratio (C/N_0) and the value of the two thresholds.

ML synchronization coherently adds the 1-ms correlated signals over each data bit interval, and then it incoherently accumulates the coherent result over several data bit intervals. Since each data bit interval has 20 PRN codes, 20 different accumulations are calculated. Each accumulation starts at a possible bit edge position. The estimated bit edge is the one that maximizes the total accumulation. In the method in [23], all of the samples must be available before it starts to process them. The method in [24] needs 20 lowpass filters, each existing at a possible bit edge position. This ML technique gives a high bit edge detection rate (EDR) if there are no phase or frequency errors, but its performance degrades in the presence of these errors.

In [25, 26], ML algorithms are introduced to estimate the bit edge position and the data bit values of a direct sequence spread spectrum (DS/SS) signal that has GPS characteristics. The algorithms assume that the phase, Doppler shift, and code delay are correctly tracked. They are intended for a small burst (3-bit) signal, and they are computationally inefficient for larger data sizes.

Methods to increase the PIT without identifying the bit edge position are discussed in [13, 14]. An overview of these methods is presented in Section 1.7. Methods to increase the PIT beyond 20 ms are proposed in [27, 28]. These methods depend on continuously estimating the data bit values and using them to remove the effect of the data bit signs. These methods assume that the bit edge position has already been identified.

In [27], two approaches are used to estimate the data bits: (1) a hard decision feedback approach, which estimates the data bits by adding the samples of each 20-ms data bit, then uses the resulting sign as the sign of that data bit, and (2) a soft decision feedback approach, which calculates a weight factor, w, for each 20-ms integrated signal. The weight factor is used as the likelihood of the signal. It is calculated from the in-phase, I, and quad-phase, Q, signals as

$$w = \begin{cases} 0 & \text{if } |I| < |Q| \\ \frac{|I|-|Q|}{|I|+|Q|} & \text{otherwise} \end{cases}$$

If $w = 0$, then the signal is too noisy to be used. If $|I| < |Q|$, then the phase error is more than 45°, so such intervals of data are not worth using in the coherent integration process. Another possibility for the weight factor is

$$w = 1 - \frac{2}{\pi}|\text{angle}\{I + jQ\}|$$

In [28], an adaptive scheme is proposed to decrease the noise variance by increasing the PIT as the C/N_0 decreases. This method also continuously estimates the data. To compensate for the effect of the bit error rate (BER), a long accumulation is used. Errors in the estimated data cause a reduction in the SNR. With perfect estimation, or without data present on the signal, the SNR is equal to

$$\text{SNR}_{NoData} = \frac{2S}{N_0\ B_p}$$

where B_p is the predetection bandwidth, and S/N_0 is the carrier-to-noise ratio, where $S/N_0 = 10^{(C/N_0)/10}$, and C/N_0 has units of dB-Hz. With nonzero BER, the SNR is reduced by a factor of

$$E[d\ d_{est}]^2 = (1 - 2\ \text{BER})^2$$

So, the SNR becomes

$$\text{SNR}_{withData} = \frac{2C\ (1 - 2\ \text{BER})^2}{N_0\ B_p}$$

This SNR can be increased by performing additional integration after the data estimation. This results in the following SNR:

$$\text{SNR}_{afterAddIntegration} = \text{SNR}_{withData} \times \text{ number of additional integrations}$$

Also, the equivalent predetection bandwidth (BW) becomes

$$B'_p = \frac{B_p}{\text{Number of additional integrations}}$$

Theoretically, if BER = 0.5, then SNR = 0; thus, no signal tracking is possible.

Methods to calculate the navigation solution without needing the navigation message are discussed in [29, 30]. In these methods, at least five satellite signals must be available.

1.9 Overview of Satellite Signal Tracking

1.9.1 Conventional GPS Tracking Loops

Following a signal acquisition, approximate code delay and Doppler shift are obtained. The acquired signal and its parameters are passed to a tracking module. The tracking module performs code, phase, and Doppler shift tracking to obtain an accurate estimate of the parameters. The phase and Doppler shift estimates are used to wipe off the carrier and, consequently, decode the navigation message. The code-delay estimate is used to calculate the pseudorange and, consequently, the navigation solution. Figure 1.3 shows a basic tracking module.

Conventional GPS tracking methods use feedback loops to extract the estimates from the noisy input measurements. A basic feedback loop consists of three main units: the mixing unit, the control unit, and the feedback unit. The mixing unit uses the noisy input measurement, z, and an internally generated expected measurement, $\hat{z}$, to generate a signal that is a function in the error of the tracked parameter. This signal is fed to the control unit, which uses a discriminator function to obtain an error signal, ϵ_p, proportional to the error in the tracked parameter. ϵ_p is fed to the feedback unit, which uses it to update the expected measurement. The updated expected measurement is fed back to the mixing unit. Thus, the tracking loop objective is to match $\hat{z}$ to z.

The expected performance of a tracking loop can be measured by the mean time to lose lock (MTLL). The MTLL is controlled by the noise variance of the

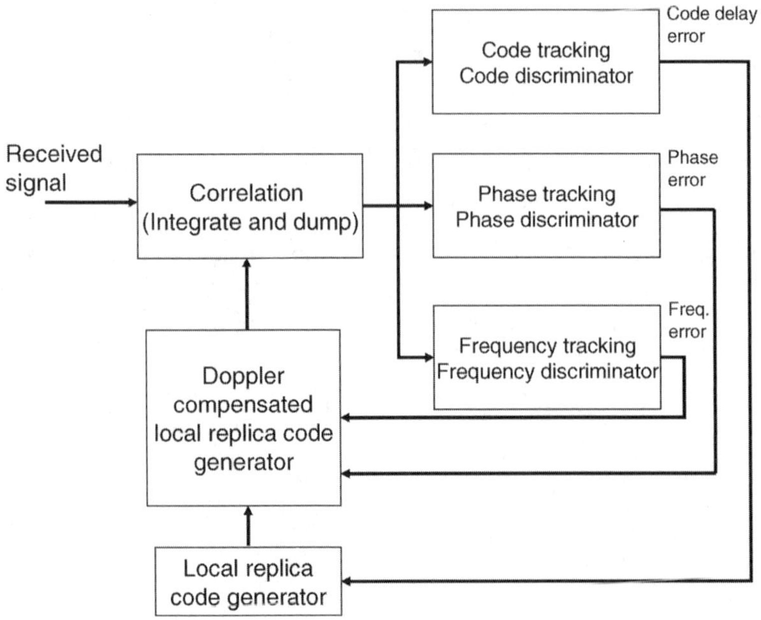

Figure 1.3 Block diagram illustrating signal tracking.

loop. The noise variance is a function in the noise bandwidth, B_n; the carrier-to-noise ratio of the input signal, C/N_0; and the integration time of the input signal, T.

There are three main types of tracking loops used in GPS receivers: delay lock loop (DLL); phase lock loop (PLL); and frequency lock loop (FLL). The GPS receiver aligns the locally generated PRN code with the received code using the estimate of the code delay, and then it generates complex local signals using the estimates of the phase and the frequency. The local signals are correlated with the received signal and used to calculate the discriminator functions of different tracking loops. Different types of signals can be generated by the correlation process. Each signal type is based on the separation between the received and the local PRN codes: prompt signals, I_{P_i} and Q_{P_i}; early signals, I_{E_i} and Q_{E_i}; and late signals, I_{L_i} and Q_{L_i}. In addition, the correlator can also generate early-minus-late signals, I_{E-L_i} and Q_{E-L_i}, where i defines the update step number. The code-delay separation between the early and late signals is defined as Δ_{EL}.

The following summarizes some of the discriminator functions used within GPS receivers. Detailed discussion about the GPS tracking loops can be found in [1–3].

1.9.1.1 Delay Lock Loop

The discriminator function produces an error signal proportional to the code-delay error, τ_{e_i}. Examples of discriminator functions and their mean and variance are as follows:

1. Noncoherent early-minus-late power discriminator:

$$\epsilon_{\tau_i} = I^2_{E_i} + Q^2_{E_i} - I^2_{L_i} - Q^2_{L_i} \tag{1.13}$$

$$E[\epsilon_{\tau_i}] \approx 2S/N_0\, T(2 - \Delta_{EL})\, \tau_{e_i} \tag{1.14}$$

$$\sigma^2_\tau = \frac{\Delta_{EL}\, B_n}{2S/N_0}\left(1 + \frac{2}{T(2 - \Delta_{EL})\, S/N_0}\right) \tag{1.15}$$

2. Noncoherent dot product discriminator:

$$\epsilon_{\tau_i} = I_{E-L_i}\, I_{P_i} + Q_{E-L_i}\, Q_{P_i} \tag{1.16}$$

$$E[\epsilon_{\tau_i}] \approx 4S/N_0\, T(1 - |\tau_{e_i}|)\, \tau_{e_i} \tag{1.17}$$

$$\sigma^2_\tau = \frac{\Delta_{EL}\, B_n}{2S/N_0}\left(1 + \frac{1}{T\, S/N_0}\right) \tag{1.18}$$

3. Coherent discriminator:

$$\epsilon_{\tau_i} = I_{E-L_i}\,\mathrm{sign}(I_{P_i}) \tag{1.19}$$

$$E[\epsilon_{\tau_i}] \approx 2\sqrt{2S/N_0\,T}\,\tau_{e_i} \tag{1.20}$$

$$\sigma_\tau^2 = \frac{\Delta_{EL}\,B_n}{2S/N_0} \tag{1.21}$$

If the PIT crosses a bit boundary, the performance will degrade.

1.9.1.2 Phase Lock Loop

The discriminator function produces an error signal proportional to the carrier phase error, θ_{e_i}. Examples of discriminator functions are as follows:

1. Costas discriminator:

$$\epsilon_{\theta_i} = Q_{P_i} I_{P_i} \tag{1.22}$$

$$E[\epsilon_{\theta_i}] \approx 2S/N_0\,T\theta_{e_i} \tag{1.23}$$

$$\sigma_\theta^2 = \frac{B_n}{S/N_0}\left(1 + \frac{1}{2T\,S/N_0}\right) \tag{1.24}$$

This discriminator is insensitive to the 180° reversal, provided that the PIT does not cross a bit boundary.

2. Decision-directed Costas discriminator:

$$\epsilon_{\theta_i} = Q_{P_i}\,\mathrm{sign}(I_{P_i}), \text{ with } 180^\circ \text{ ambiguity} \tag{1.25}$$

3. Arctangent discriminator:

$$\epsilon_{\theta_i} = \arctan\frac{Q_{P_i}}{I_{P_i}} \tag{1.26}$$

The sign changes can be detected and used, provided that the PIT does not exceed one data bit interval or cross a bit boundary; however, there is a 180° phase ambiguity, which can be resolved during the frame synchronization.

1.9.1.3 Frequency Lock Loop

The discriminator function produces an error signal proportional to the change in the carrier phase error over a finite time interval. Examples of discriminator functions are as follows:

1. Cross-product discriminator:

$$\epsilon_{f_i} = I_{P_{i-1}} Q_{P_i} - I_{P_i} Q_{P_{i-1}} \tag{1.27}$$

If the $i-1$ and i signals belong to different data bits with different signs, then the discriminator produces the wrong sign.

2. Decision-directed cross-product discriminator:

$$\epsilon_{f_i} = (I_{P_{i-1}} Q_{P_i} - I_{P_i} Q_{P_{i-1}}) \operatorname{sign}(I_{P_{i-1}} I_{P_i} + Q_{P_{i-1}} Q_{P_i}) \tag{1.28}$$

This discriminator modulates the cross product with the sign of the dot product. This means that ϵ_{f_i} will have the correct sign if the $i-1$ and i signals belong to different bits with different signs. The sign changes can be used to detect the data bits.

3. Differential arctangent discriminator:

$$\epsilon_{f_i} = \arctan \frac{Q_{P_i}}{I_{P_i}} - \arctan \frac{Q_{P_{i-1}}}{I_{P_{i-1}}} \tag{1.29}$$

If the $i-1$ and i signals belong to different data bits, then ϵ_{f_i} might have the wrong sign, depending on the quadrant of the signals. FLL will not have frequency ambiguity because of data bit transition provided that adjacent I and Q signals are taken within the same data bit interval. The sign changes can be detected, but there might be sign ambiguity, which can be resolved during frame synchronization.

1.9.2 Satellite Signal-Tracking Techniques

There are two main tracking approaches: hardware-based and software-based tracking. Hardware tracking uses conventional tracking loops [1, 2]. Software tracking implements all of the functionalities of tracking loops in software [5]. This gives the software-based tracking approach more flexibility to use better tracking techniques and, consequently, enables better performance. In addition, software-based tracking approaches can perform non-real-time tracking. There have been few tracking approaches designed for weak signals or for software receivers [31–34]. Such approaches could enable the use of GPS in wireless applications with less cost and more accuracy.

In [31], techniques are introduced to enable the tracking of weak sidelobe signals in high-altitude satellites. A fine acquisition algorithm is developed to improve the estimation accuracy of the Doppler shift and code delay, which are obtained using an acquisition process. The fine acquisition algorithm also estimates the bit edge position, the carrier amplitude, the phase, and the rate of change of the Doppler shift. The output of the fine acquisition is used to initialize a combined carrier- and code-tracking algorithm, which is based on EKF approaches.

The fine acquisition algorithm first estimates the bit edge position using an ML approach. Then, it builds a cost function of the unknown phase, Doppler shift, Doppler rate, signal amplitude, and code delay. These unknowns can be estimated by a minimization operation. The algorithm estimates most of the parameters by reducing the original cost function to simpler cost functions, which are easier to solve. The EKF tracking algorithm is implemented to use two assumptions, the known and unknown data bit values. If the data bits are known, the filter performs one measurement update and state propagation over each data bit by minimizing a cost function. Then, it propagates the solution to the next step. If the data bits are unknown, a Bayesian estimation is used, where the EKF calculations are done twice, once for each possible data bit value. The a posteriori probabilities of the possible data bit values (i.e., +1 and −1) are derived using the conditional probability density for the state, given the measurements and the bit sign assumptions. The bit sign assumption that gives the smallest optimal cost will have the highest a posteriori probability. These probabilities are multiplied by the corresponding propagated states. The results are added to produce a mixed state update and a mixed covariance update. The mixed covariance function also includes terms that increase the covariance update in the direction of the correct bit value estimate.

In [32], another tracking algorithm for weak signals is developed. The algorithm assumes that the Doppler shift has been acquired at a resolution of 1 Hz, the bit edge is known, and the receiver is stationary. A PIT of 1 sec. is used. It is assumed that the maximum change of the Doppler shift over 1 sec. is less than 1 Hz. First, a biphase approach is used to detect the navigation data. Then, the algorithm deals with two problems: the alignment of the local code with the received one, and the change in the chip length because of the Doppler effect.

For each 1 sec. of data, 50 points are produced for each of the early, prompt, and late signals. Each point corresponds to one data bit interval. The prompt points are used to decode the data. This is done by dividing the 50 points into 25 pairs and comparing the sum and the difference of each pair. This process aims to detect the existence of a data transition. The data transition is removed by multiplying the two points by 1 and −1, then adding them together.

The resultant phase is compared to a reference phase. If the phase difference is larger than 90°, the two points are multiplied by −1.

Following the data decoding, each 50 points of the three signals are added together. Then, a simple method is used to find the code-delay error in which the autocorrelation function is assumed to have a triangular shape. The frequency offset is found by applying a 50-point FFT to the 1-sec. data, and then the amplitude of the peak is compared to the amplitudes of its two neighbors.

The local code is not generated with a length depending on the Doppler shift. Instead, the change in the code duration is deduced and used either to remove the amount of the code change from the calculated code delay or to generate the local code at a delay equal to the amount of that change. The result shows the ability of this method to track a signal down to 18 dB-Hz, with a standard deviation around 45 ns.

In [35], a tracking algorithm is developed based on the Kalman filter (KF) and the maximum likelihood (ML) approaches. It aims to enable tracking in the presence of false modulation on the phase and amplitude. Such modulation causes the existence of unknown parameters, which are needed to estimate the states of the KF correctly. The ML is employed to maximize the log likelihood of the measurement matrix. The likelihood function is extended over a time interval, and it is conditioned on the unknown parameters. Two KF are used: a reference KF for the estimation of the unknown parameters and a primary KF for the state estimation.

In [33], a tracking algorithm based on KF smoothing is developed. The algorithm is designed to work with non-real-time data of signals with high C/N_0. Its objective is to provide carrier phase and code tracking with good accuracy; this is done through the use of future data relative to the estimation time. The tracking starts by obtaining estimates of the tracked values using a KF. The KF is implemented in such a way as to resemble the function of a phase lock loop (PLL) and a delay lock loop (DLL). The phase of the carrier and code are estimated using discriminator functions. The outputs of the discriminators are used to calculate the KF innovations, which represent the difference between the discriminators' outputs and the propagated values. The PLL is then realized through a steady-state version of the KF. The KF gain is assumed to approach a constant value as time goes to infinity, and this value is used in the PLL. A similar approach is used for the DLL realization.

The output of the KF is used in a smoothing algorithm that processes data in batches. Each batch of data is extended over a fixed interval. A refined state estimate is produced based on the entire batch. The smoothing problem is formulated as a least-square problem. It aims to find the optimal time histories for the state, the process noise, and the measurement noise. The smoothing problem is solved by using a modified form of the standard square root information filter/covariance smoother algorithm. The algorithm goes through the data of each batch twice, once in a forward-direction and once in a backward-direction. The forward-direction process is similar to the KF, while the backward-direction process is a recursive operation.

A tracking algorithm, block adjustment of synchronizing signal (BASS), is developed in [34]. It is a software approach that first resolves the phase and the frequency to a fine resolution using an FFT approach. Then, it resolves the code-delay error using a method that assumes that the autocorrelation function has a triangular shape. The fine phase and frequency are calculated as follows. FFT is calculated over N received samples. An approximate estimate of the frequency is taken as the bin that produces the peak amplitude; the phase estimate is deduced from the real and imaginary parts of that bin. Then, a refined frequency estimate, f, is calculated from the calculated phase values at two different times, θ_n and θ_{n+m}, separated by the processing time, m, as

$$f = \frac{\theta_{n+m} - \theta_n}{m} \tag{1.30}$$

Following that, a replica code is generated and used throughout the tracking process.

A tracking technique that combines the tracking loops with KF approaches is developed in [36]. It uses two different types of PLL, one for the tracking of the receiver dynamics and another for the tracking of the satellite dynamics. Only one common PLL of the first type is used, and N PLLs of the second type are used, one for each tracked satellite. The idea is to use the common PLL to provide a performance enhancement to the other PLLs. A vector containing the discriminator outputs of the N PLLs is transformed into a four-dimensional vector, ϵ, using a covariance matrix similar to the one that is usually used in the KF, where ϵ is a tracking-error vector. ϵ is used in the common PLL; it is filtered to produce the position and time errors, and then it is transformed into an N-dimensional vector, Z_g, using a matrix that depends on the relative positions of the user and the satellites. Each of the N PLLs has a numerically controlled oscillator, which is controlled using a signal obtained from the summation of three vectors. The first vector is Z_g, the second consists of the predicted values of the Doppler shifts, and the third is a signal obtained from the filtered output of each of the N PLLs.

Software approaches for processing GPS signals are discussed in [37]. The objective is to obtain a software receiver that can be used in wireless phones. The signal is sampled directly near the antenna and passed to the software module. For the acquisition process, a Doppler-compensated replica code is generated and used to calculate the integration by circular correlation. Interpolation is used to reduce the code-delay error. For the tracking process, a first-order DLL and a second-order FLL are used.

The transition from acquisition to tracking is discussed in [38]. Since the start of the C/A code phase changes with time because of the Doppler shift effect, the objective is to obtain an estimate of the start of the code at the beginning of the tracking time. The receiver is assumed to be stationary, so the relationship between the Doppler shift and the rate of change of the C/A code delay is linear. This relationship is used to obtain a relationship between the cumulative delay shift and time. The start of the code is estimated using the Doppler shift and the clock error. This is done using the following equation:

$$pt = pt_0 + \left(-\frac{f_s}{f_c} T_{C/A} f_d + err \right) t$$

where pt is the estimated C/A code start sample; pt_0 is the initial C/A code start sample (obtained from the acquisition); f_s, f_c, and f_d are the sampling frequency, carrier frequency, and Doppler shift, respectively; $T_{C/A}$ is the C/A code length; and *err* is the clock error expressed in units of samples per second. This method can be applied if the transition time from acquisition to tracking is relatively short; however, if the code-delay error exceeds half a chip, the tracking module will not lock onto a satellite. Thus, the cumulative error in the code delay, resulting from the clock error, plus errors in the analog-to-digital converter, the inaccurate Doppler shift, and the change in the Doppler shift, must be less than half a chip during the transition time.

1.10 Navigation Message Decoding

The navigation message (NAV) of the GPS L1 C/A signal is encoded by (32,26) Hamming code [4], where the 32 data bits represent one word, plus the last two bits of the previous word. Bits 25 to 30 of each word contain parity bits. The encoding described in [4] is as follows: $d_1, d_2, \ldots, d_{30}$ are the source data bits; $D_1, D_2, \ldots, D_{30}$ are the encoded data bits, which are transmitted by the satellite; D_{29}^* and D_{30}^* are the last two bits of the previous word. The encoding of d_i, for $i = 1, \ldots, 24$, is done as

$$D_i = d_i \oplus D_{30}^* \tag{1.31}$$

where $\oplus$ defines the exclusive-or operation. The parity bits are calculated as

$$\begin{aligned}
D_{25} &= D_{29}^* \oplus d_1 \oplus d_2 \oplus d_3 \oplus d_5 \oplus d_6 \oplus d_{10} \oplus d_{11} \oplus d_{12} \oplus d_{13} \oplus d_{14} \oplus \\
&\quad d_{17} \oplus d_{18} \oplus d_{20} \oplus d_{23} \\
D_{26} &= D_{30}^* \oplus d_2 \oplus d_3 \oplus d_4 \oplus d_6 \oplus d_7 \oplus d_{11} \oplus d_{12} \oplus d_{13} \oplus d_{14} \oplus d_{15} \oplus \\
&\quad d_{18} \oplus d_{19} \oplus d_{21} \oplus d_{24} \\
D_{27} &= D_{29}^* \oplus d_1 \oplus d_3 \oplus d_4 \oplus d_5 \oplus d_7 \oplus d_8 \oplus d_{12} \oplus d_{13} \oplus d_{14} \oplus d_{15} \oplus \\
&\quad d_{16} \oplus d_{19} \oplus d_{20} \oplus d_{22}
\end{aligned}$$

$$D_{28} = D_{30}^{*} \oplus d_2 \oplus d_4 \oplus d_5 \oplus d_6 \oplus d_8 \oplus d_9 \oplus d_{13} \oplus d_{14} \oplus d_{15} \oplus d_{16} \oplus d_{17} \oplus d_{20} \oplus d_{21} \oplus d_{23}$$

$$D_{29} = D_{30}^{*} \oplus d_1 \oplus d_3 \oplus d_5 \oplus d_6 \oplus d_7 \oplus d_9 \oplus d_{10} \oplus d_{14} \oplus d_{15} \oplus d_{16} \oplus d_{17} \oplus d_{18} \oplus d_{21} \oplus d_{22} \oplus d_{24}$$

$$D_{30} = D_{29}^{*} \oplus d_3 \oplus d_5 \oplus d_6 \oplus d_8 \oplus d_9 \oplus d_{10} \oplus d_{11} \oplus d_{13} \oplus d_{15} \oplus d_{19} \oplus d_{22} \oplus d_{23} \oplus d_{24}$$

At the receiver, the data are decoded as follows [4]:

(1) If $D_{30}^{*} = 1$, go to step (2a); otherwise, go to step (2b).

(2a) Complement $D_1, \ldots, D_{24}$ to get $d_1, \ldots, d_{24}$. Go to step (3).

(2b) $d_1, \ldots, d_{24}$ are taken equal to $D_1, \ldots, D_{24}$.

(3) Substitute $d_1, \ldots, d_{24}$ into the parity equations to get computed parity bits.

(4) If the computed parity bits are equal to the corresponding received parity bits, go to step (5a); otherwise, go to step (5b).

(5a) Parity check passes. Conclude the decoded data for that word as $d_1, \ldots, d_{24}$.

(5b) Parity check fails. No decoding is concluded for that word.

1.11 GPS Modernization and Galileo Global Navigation Satellite System

1.11.1 GPS Modernization

There has been a large growth in the applications that employ GNSS for positioning. The current operational GNSSs, GPS and GLONASS, were not designed for such applications. The civil GPS C/A signal on the L1 carrier frequency suffers from many limitations. It has a short code length, 1,023 chips, that makes the signal vulnerable to cross correlation and interference problems. Its transmission rate, 1.023 Mchip/s, limits the position accuracy that can be achieved. The capability of the positioning degrades badly in environments like urban areas and inside buildings due to multipath and signal attenuation. The simple encoding of the navigation message and the parity check do not allow for any corrections to the received data bits. This limits the decoding capability and, consequently, the minimum BER that can be achieved for a C/N_0.

The limitations of the current GPS civil signal and the new wide range of applications have motivated the design of new GPS signals with enhanced capabilities. The new signals will offer new modulations, new signal structures, longer codes, dataless codes, faster transmission rates, new data encoding, a new navigation message format, and so on.

The first modernized GPS signal is the L2C civil signal on the L2 carrier frequency. The first Block IIR-M GPS satellite carrying this signal was launched in late September 2005. The L2C signal consists of two codes, CM and CL, that are multiplexed chip by chip. Figure 1.4 illustrates the generation of the multiplexed CM and CL codes. Each code is generated with a rate of 511.5 KHz. The two codes are combined before transmission such that a chip of the CM code is transmitted followed by a chip of the CL code. Consequently, the chipping rate is 1.023 MHz, which is the same as that of the C/A code. The CM code has a length of 10,230 chips and repeats every 20 ms; it is modulated by a 50-Hz message. The CL code is dataless, it is 75 times longer than the CM code and has a length of 767,250 chips. The CL code repeats every 1.5 sec. and is synchronized with the Z-count. The Block IIR-M satellites will transmit the navigation message with the same format used with the C/A signal. The message can be convolutionally encoded with a rate of 1/2 and a constraint length of 7.

The shift registers used to generate the CM and CL codes have 27 stages, with 12 feedback tapes. Such logic can produce a code with a maximum number of chips equal to $2^{27} - 1 = 134, 217, 727$. However, both codes are produced by initializing the shift register to a predefined initial state and continuing with the code generation until either the required number of chips is obtained or the final

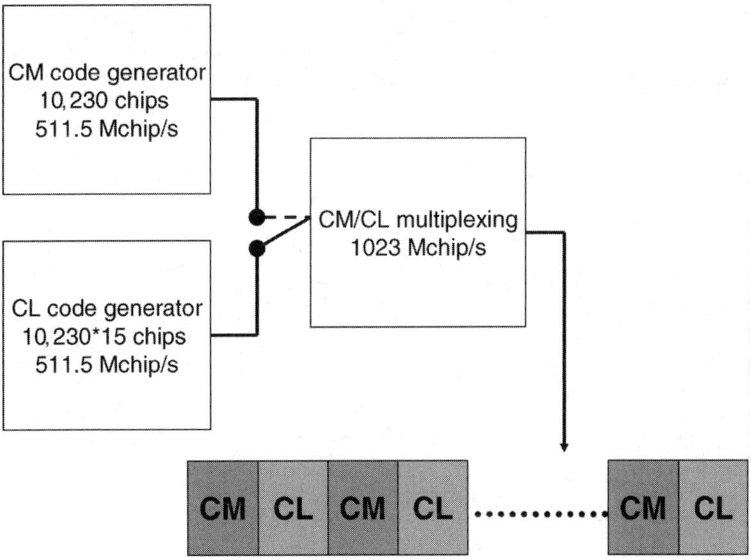

Figure 1.4 The generation of the multiplexed CM and CL codes for the L2C signal.

state is detected. Following that, the register is restored to the predefined initial state. Both of the codes are perfectly balanced (i.e., the number of zeros is equal to the number of ones). The multipath performance can be equivalent to, or better than, a continuous code with a chipping rate of 10.230 MHz.

In addition, two new civil signals will be introduced on the L5 carrier frequency. The L5 carrier frequency is 1,176.45 MHz. The signals will be transmitted by the Block IIF satellites. There will not be shared military signals. The L5 carrier frequency is modulated by a quadrature phase shift keying (QPSK), with the two civil signals on the inphase and quadphase channels, I5 and Q5. The two signals will have equal power. The L5 signal is about 3 to 4 dB stronger than the L1 or L2 signals. The PRN codes of the I5 and Q5 channels are different, and each code has a length of 10,230 chips and a transmission rate of 10.23 MHz. Thus, each code has a 1-ms repetition. The codes are generated by performing an Exclusive-Or operation using two 13-stage shift registers. The I5 signal is modulated by a 100-Hz navigation message, which is encoded by a forward error correction (FEC) and a 10-bit Neuman-Hoffman code. Figure 1.5 shows the timing relationship between the PRN code, the Newman-Hoffman code, and the data bits. The Q5 signal is dataless and modulated by a 20-bit Neuman-Hoffman code. The Block IIF will transmit the navigation message in a new format (CNAV); it will be convolutionally encoded.

1.11.2 Galileo European Satellite Navigation System

Galileo is a new GNSS financed by Europe. It has been developed collaboratively by the European Union (EU), the European Space Agency (ESA), and European industry. It is designed for civilian applications, but it is also aimed to be used for security applications. Galileo is to provide enhanced signal structure and system capabilities in order to allow for the availability of new services and applications. Galileo services include (1) open service (OS), free of charge, to be

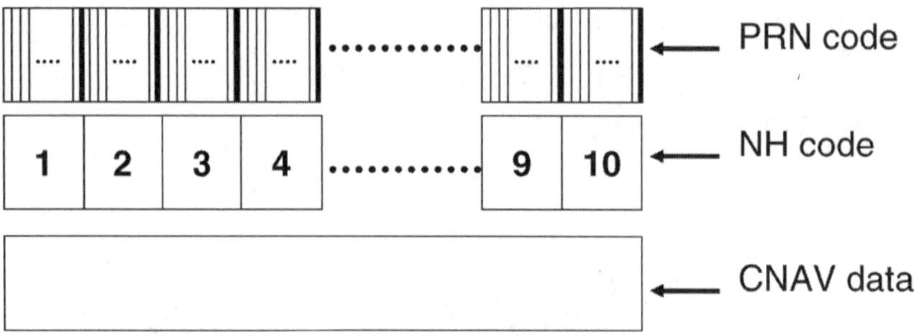

Figure 1.5 Timing relationship between the PRN code, the Newman-Huffman code, and the navigation data bits of the L5 GPS signal.

used for mass-market applications; (2) commercial service (CS) for applications with performance requirements that exceed those of the OS; (3) safety of life (SOL) service for transport applications; (4) public regulated service (PRS) for governmental applications; and (5) search-and-rescue (SAR) support service.

The Galileo satellite constellation will consist of 30 satellites distributed in three orbital planes inclined by 56° with respect to the equator and located at an altitude of about 23,616 km (i.e., in medium earth orbit [MEO]). Each orbit is to have three spare slots for redundant satellites.

The navigation concept of Galileo is similar to that of GPS. Galileo, however, is intended to provide high accuracy and guaranteed positioning and to be under civilian control. In addition, Galileo is to be interoperable with existing GNSSs.

Galileo signals are to be transmitted on three frequency bands using four carrier frequencies [39, 40]:

1. E5 band with center frequency of 1,191.795 MHz. Two carrier frequencies will be transmitted on the E5 band: E5a with a carrier frequency of 1,176.45 MHz and E5b with a carrier frequency of 1,207.14 MHz.
2. E6 band with a center frequency of 1,278.75 MHz.
3. L1 band with a center frequency of 1,575.42 MHz.

Two types of subcarrier modulation are to be used: BPSK and binary offset carrier (BOC). There are a total of 10 signals, which are as follows:

- *Two signals on the E5a frequency (i.e., on I and Q channels) and two signals on the E5b frequency.* One possibility is that the E5a and E5b frequencies will be modulated by QPSK. Each of the four signals is modulated by BPSK. The codes have a chipping rate of 10.23 Mchip/s. The signal on the E5a/I channel is modulated by a 50 symbol/s (25 bps) navigation message. The signal on the E5b/I channel is modulated by a 250 symbol/s (125 bps) navigation message. The E5a/Q and E5b/Q channels are dataless pilot signals.
- *Three signals on the E6 frequency, A, B, and C channels.* The E6 carrier is to be modulated by coherent adaptive subcarrier modulation (CASM). The signal on the A channel is modulated by BOC; the signals on the B and C channels are modulated by BPSK. The codes of the three signals have a chipping rate of 5.115 Mchip/s. The signal on the A channel is modulated by a 250 symbol/s (125 bps) navigation message; the signal on the B channel is modulated by a 1,000 symbol/s (500 bps) navigation message; the signal on the C channel is a dataless pilot signal.

- *Three signals on the L1 frequency, A, B, and C channels.* The L1 carrier is to be modulated by CASM. Each of the three signals is modulated by BOC. The code of the A channel has a chipping rate of 5.115 Mchip/s; the codes of the B and C channels have a chipping rate of 1.023 Mchip/s. The signals on the A and B channels are modulated by a 250 symbol/s (125 bps) navigation message; the signal on the C channel is a dataless pilot signal.

1.12 Weak-Signal Applications

A GNSS receiver must be able to work under weak-signal and various dynamic conditions in some applications, for example, to provide a positioning capability in wireless devices or orbit determination for geostationary and high-Earth-orbit satellites. The following subsections discuss some weak-signal applications.

1.12.1 Wireless and Indoor Positioning

Three approaches can be used in the positioning of wireless devices [41–43]: network-based positioning, stand-alone GPS, and assisted GPS (AGPS).

The network-based-positioning approach does not use GPS technology. Instead, the user position is calculated by the network itself. This can be done using different methods, such as time of arrival and angle of arrival. The user position is calculated based on measurements taken at different base stations. If the time-of-arrival method is used, then the user position is determined based on the difference between the signal arrival times at different base stations. Network-based positioning can determine the user position in a shorter time compared to GPS positioning (i.e., shorter time to first fix [TTFF]), and it does not require that any complexity be added to the wireless devices; however, it has high cost for the base stations and is less accurate than GPS positioning. It also has some privacy issues because users have no control over their tracking by the network. Therefore, there has been an interest in using GPS in the positioning of wireless devices.

The problem with using a stand-alone GPS receiver for the positioning of wireless devices is that these devices receive signals with very low power, typically below a C/N_0 of 25 dB-Hz, due to the attenuation incurred from the surrounding environment, for instance when operating indoors or under heavy vegetation. Complex algorithms are required to acquire and track weak signals. Usually, such algorithms have high processing and memory requirements, but wireless devices usually have small processors, small memory size, and limited battery life.

The approach that combines the benefits of both network-based positioning and GPS positioning is AGPS. AGPS is a technology that can be employed within a wireless network. It consists of the following three components: (1) a wireless device that contains a partial GPS receiver, (2) an AGPS server with a reference GPS receiver that can see the same satellite as the wireless device, and (3) a wireless network infrastructure. The network sends the wireless device a prediction of the received GPS signal. This could include approximate Doppler shift, code delay, and the navigation message. This enables the use of a long coherent integration time, which enhances sensitivity and reduces the search space of the acquisition process. Since there is information sharing between the user and the network, the user can preserve privacy by controlling what is being sent, and the network can restrict the information shared by some users. In [42], an overview is provided of the application of GPS technology to wireless devices; the benefits of using AGPS for indoor positioning are also discussed. In [43], a more detailed discussion is provided of the AGPS architecture and of the interaction between the wireless devices and the network.

The AGPS technology provides assisting information, including the ephemeris, to the GPS receiver. Usually, the ephemeris is valid for up to 4 hours only, so there has been a shift in the AGPS technology to synthesize the satellite orbits. This can allow for the generation of predicted satellite orbits that can be valid for several days.

An example of an AGPS is a system developed by a company called Global Locate [44–51]. This AGPS has a worldwide network of reference stations that track all GPS satellites at all times. This system has two main features. First, the hardware chip that must be added to the wireless devices has a massive number of correlators to search all of the code delays using long integration time. Second, the system generates an orbit model that is valid for up to 10 days. The system can perform the position calculation at the device or at a network server using two modes: (1) mobile station (MS)–assisted mode, where the position calculation is done at the server, and (2) MS-based mode, where the position calculation is done at the wireless device. Another example of an AGPS is a system developed by a company called snapTrack [52–58].

In general, to take advantage of the AGPS technology, special algorithms need to be designed. The AGPS, however, is not always available, as is the case in sensor networks [59]. A sensor network consists of a large number of small devices that are deployed in some areas to monitor and detect physical events. Accurate positioning is needed in such networks for efficient functionality and resource coordination. The devices' locations can change arbitrary, and their batteries, which determine their lifetimes, are limited. Poor GPS performance in the deployment areas and the unavailability of efficient GPS receiver algorithms have been the main obstacles limiting the use of GPS technology [60, 61]. In such applications, the problem of the cost can be mitigated by installing GPS receivers in a small number of the devices and designing algorithms to handle communications between different devices.

1.12.2 Orbit Determination of Geostationary and High-Earth-Orbit Satellites

Geostationary and high-Earth-orbit (HEO) satellites are located above the GPS constellation. Figure 1.6 illustrates the location of HEO with respect to the GPS constellation and Earth. The GPS antenna is oriented toward the Earth, so the transmitted signals are not accessible to any receiver above the transmitting GPS satellite. However, the antenna pattern enables those signals to span the space beyond the Earth and to become visible to receivers on the opposite side of the Earth. However, the GPS signals are received with very low power due to the high path loss of the transmitted signals. In addition, the number of visible satellites is small, and each satellite is only available for short periods.

1.13 Overview of Some Technical Concepts

1.13.1 Viterbi Algorithm

The Viterbi algorithm (VA) is a technique that has been used in many applications. Its objective is to find the optimal solution to a problem, with optimal processing. It estimates the most likely sequence of data, given a set of observations. This is done by dynamic programming, which is an optimal recursive search process [62]. In general, dynamic programming can be used to solve a problem if it has the following properties: (1) an optimal structure (i.e., the optimal solution to the problem consists of optimal solutions to its subproblems), and (2) the optimal solution can be expressed in a recursive form. The idea behind the VA is that if the most likely sequence of data between points P_0 and P_N passes through point P_i, then the subsequence from P_0 to P_i is also the most likely subsequence.

Typically, a problem is formulated for the VA by representing the system in question with a state diagram, then deriving a trellis graph from it. A state diagram of a system with M states that are fully connected has M^2 transitions.

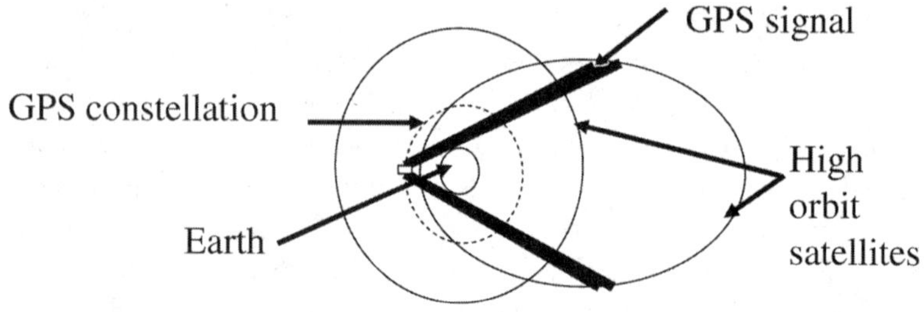

Figure 1.6 HEO satellites' location with respect to the earth and the GPS constellation.

If there are N observations, then there are M^N possible sequences. Each sequence is represented by a path in the trellis graph, which keeps track of all of the possible transitions through the N steps. The optimal sequence can be found by associating a weight with each transition and finding the path with the optimal (minimum or maximum) total weight. Although the optimal sequence can be found by searching all of the paths, such a process is impractical. Instead, the VA operates recursively and keeps track only of the most likely subsequences after processing each observation. If there is more than one path to the same state at step i, then only the survivor path, the path with the optimal total weight up to this step, is retained. After processing each observation, there will be M or fewer survivor paths. After processing all of the observations, the survivor path with the optimal total weight will contain the most likely sequence.

For example, the VA is used for symbol sequence, channel, and/or phase estimation [63–65]. In [63], VA paths are generated for each possible phase value. While this technique provides optimal estimation, it is computationally inefficient. A review of some methods that reduce the computation is presented in [64]. Generally, the phase is estimated for each survivor path separately. In [64, 65] phase tracking is done using a second-order loop attached to each survivor path. The estimated phase error is used to counterrotate the next received signal with that error before the VA processes it.

1.13.2 Estimation Techniques

Estimation techniques are classified into two main approaches: Bayes (e.g., maximum a posteriori [MAP]) and classical (e.g., maximum likelihood [ML]).

Let $\mathbf{r}$ define a random variable with a probability density function (PDF) of $f_r(\mathbf{r}, \alpha)$. r defines the observations; $r = (r_1, r_2, \ldots, r_n)$. α defines the parameters that need to be estimated; $\alpha = (\alpha_1, \alpha_2, \ldots, \alpha_m)$. Assuming independent observations with the same distribution, the likelihood function of $\mathbf{r}$ is the joint PDF of n observations, r_k. This function is

$$f_r(\mathbf{r}, \alpha) = \prod_{k=1}^{n} f_r(r_k, \alpha) \tag{1.32}$$

An unbiased estimate satisfies [66]

$$E[\hat{\alpha}_i] = \alpha_i \tag{1.33}$$

$$\sigma_{\alpha_i}^2 = \mathrm{var}[\alpha_i - \hat{\alpha}_i] \geq J_{ii} \tag{1.34}$$

where $\hat{\alpha}_i$ is an estimated value of α_i; $E[.]$ is the mean function; and $\sigma^2_{\alpha_i}$ is the variance. J is the Fischer's information matrix with a size of $m \times m$, where

$$J_{ij} = -E\left[\frac{\partial^2 \ln f_r(\mathbf{r}|\alpha)}{\partial\alpha_i\,\partial\alpha_j}\right] \tag{1.35}$$

1.13.2.1 MAP Estimation

In MAP estimation, the estimated parameters, α, are considered random variables with known PDFs, f_{α_i}. The MAP technique searches for the $\hat{\alpha}_i$ value that maximizes the a posteriori PDF; that is,

$$\hat{\alpha}_i = \arg\max_{\alpha_i} f_{\alpha_i|r} = \arg\max_{\alpha_i} \frac{f_{r|\alpha_i}\, f_{\alpha_i}}{f_r} \tag{1.36}$$

1.13.2.2 ML Estimation

If the PDFs of the estimated parameters are not available, then ML estimation is used. The estimated parameters are assumed to be uniformly distributed. The ML technique searches for the $\hat{\alpha}_i$ value that maximizes the likelihood function; that is,

$$\hat{\alpha}_i = \arg\max_{\alpha_i} f_{r|\alpha_i} \tag{1.37}$$

Since this function is conditioned on the estimated parameter, the maximized function depends only on the noise distribution.

1.13.3 Kalman Filter

The Kalman filter (KF) was first introduced in [67]. The KF provides a recursive solution to linear problems. Define z_i as the measured observation at time instance t_i, s_i as an unknown signal, and n_i as an unknown measurement noise:

$$z_i = s_i + n_i \tag{1.38}$$

The KF is formulated to solve the following problem. Given the observations $z_0, \ldots, z_i$, estimate the signal parameters at time t_m. If $t_m < t_i$, the KF solves a data-smoothing problem. If $t_m = t_i$, the KF solves a filtering problem. If $t_m > t_i$, the KF solves a prediction problem.

A complete introduction to and analysis of the KF is introduced in [68, 69]. The state propagation is described by a linear stochastic differential equation of the form

$$\dot{x}(t) = F(t)\, x(t) + B(t)\, u(t) + G(t)\, w(t) \tag{1.39}$$

where x is a state process vector with size v; F is a system dynamic matrix of size $v \times v$; u is a vector of control input functions; B is a deterministic input matrix; w is a vector of white Gaussian process noise with zero mean and Q covariance matrix; and G is a noise input matrix. The observations, or measurements, are taken at discrete time instances. The measurement vector is

$$z_i = H_i \, x_i + n_i \tag{1.40}$$

where H is a measurement matrix that linearly relates z to the state vector x, and n is a vector of white Gaussian measurement noise with zero mean and R covariance matrix. The KF provides models for the uncertainties introduced by different noises. It propagates the state estimate at each step, taking into account the system dynamics and the time history of the state estimate; thus, an expected measurement is calculated. It achieves optimal estimation by combining the actual and the expected measurements and taking into account the measurement noise and the estimated uncertainty in its current state.

The KF operation can be summarized as follows. From the state propagation description in (1.39), assuming that the control input u is zero, the system dynamic model is found as

$$x_{i+1} = \Phi \, x_i + w_i \tag{1.41}$$

where Φ is the transition matrix, which relates the states at step $i + 1$ to the states at step i. It is derived from Taylor series as

$$\Phi = \mathbf{I} + F t + \frac{1}{2} F^2 t^2$$

where $\mathbf{I}$ is an identity matrix. Following that, the continuous time t is replaced with the measurement interval to fit the discrete measurements. Let x_i^- define the a priori, or propagated, state estimate at step i. x_i^+ defines the a posteriori, or updated, state estimate at step i. P_i^+ defines the updated error covariance matrix. P_i^- defines the propagated error covariance matrix.

The KF works recursively. In each KF step, there are two stages: (1) updating the state estimate and the error covariance matrix using the new measurement z_i, and (2) propagating the error covariance matrix and the state estimate to the next step. The KF operation starts by initializing x_0^- and P_0^-. Each step starts by calculating the KF residual, or innovation, as the difference between the actual measurement and the expected measurement (i.e., $z_i - H_i \, x_i^-$). A gain, K_i, is calculated to weigh the residual so as to minimize the covariance of the error in the updated estimate (i.e., to minimize $E[(x_i - x_i^+)\,(x_i - x_i^+)^T]$). The gain is calculated from

$$K_i = P_i^- \, H^T \, (H \, P_i^- \, H^T + R)^{-1} \tag{1.42}$$

The updated estimate is obtained as

$$x_i^+ = x_i^- + K_i \left(z_i - H x_i^- \right) \tag{1.43}$$

The error covariance matrix is updated as

$$P_i^+ = (\mathbf{I} - K_i\, H_i)\, P_i^- \tag{1.44}$$

Following that, the state vector and the error covariance matrix are propagated to the next step as follows:

$$x_{i+1}^- = \Phi\, x_i^+ \tag{1.45}$$

$$P_{i+1}^- = \Phi\, P_i^+\, \Phi^T + Q_{i+1} \tag{1.46}$$

There are systems where the measurements or the dynamics equations are nonlinear. The extended Kalman filter (EKF) is used for estimation in such systems. The EKF linearizes around the current estimate using Taylor series. The state propagation in (1.39) and the dynamics model in (1.41) are modified as

$$\dot{x}(t) = f(x(t), u(t)) + G(t)\, w(t) \tag{1.47}$$

$$x_{i+1} = f(x_i) + w_i \tag{1.48}$$

The measurement vector in (1.40) is modified as

$$z_i = h(x_i) + n_i \tag{1.49}$$

The linearization is done to obtain the Φ matrix and the H matrix. The elements of these matrices are obtained as

$$\Phi_{i,\,jk} = \frac{\partial f_j}{\partial x_k}(x_i^+) \tag{1.50}$$

$$H_{i,\,jk} = \frac{\partial h_j}{\partial x_k}(x_i^-) \tag{1.51}$$

where each of $\Phi_{i,\,jk}$ and $H_{i,\,jk}$ refers to the element at the intersection of the jth row and kth column; f_j and h_j refer to the jth element; and x_k refers to the kth element of the state vector. The expected measurement is obtained as

$$\hat{z}_i = h(x_i^-) \approx H_i\, x_i^- \tag{1.52}$$

The updated estimate is obtained as

$$x_i^+ = x_i^- + K_i\,(z_i - \hat{z}_i) \tag{1.53}$$

The rest of the equations are the same as those used with the linear case.

1.14 GNSS Positioning Using Weak Signals

1.14.1 Weak-Signal Positioning Problems

Different PRN codes have small cross correlation, and the same PRN codes correlate if their relative delay, τ, is less than one chip length, T_{chip}. The autocorrelation is proportional to $(1 - |\tau|/T_{chip})$. If the codes are perfectly aligned with each other, the autocorrelation produces a maximum value. In order for a GPS satellite signal to be acquired and tracked, the receiver generates a replica PRN code for that satellite and correlates it with the incoming signal. The correlated signals are then accumulated coherently over T_I ms, which defines the PIT; the result can be further accumulated incoherently. The accumulated signal is used for signal acquisition and tracking.

Acquisition of very weak signals requires long coherent and incoherent integrations. The coherent integration decreases the noise bandwidth and improves sensitivity, but it is limited by the unknown data bits and bit edges. In the case of the GPS C/A code, the upper bound on the coherent integration is one code length (i.e., 1 ms) if the data bits and the bit edge positions are unknown, and it is one data bit length (i.e., 20 ms) if only the data bits are unknown. The increase in the coherent integration causes an increase in the number of Doppler bins that need to be searched. Thus, it increases the processing and memory requirements. On the other hand, the incoherent integration averages amplitudes, so it does not suffer from similar limitations. It is, however, less sensitive because of the squaring loss, and it becomes less effective as its number increases.

High dynamic acquisition introduces another problem. The Doppler shift continues to appear on different bins with a rate depending on the Doppler rate. Therefore, long integration can scatter the signal power over many bins. This puts a limit on the maximum integration length that can be used and, consequently, on the minimum acquired signal strength.

Although different PRN codes have small cross-correlation, the cross-correlation of a strong signal can result in a stronger power than the autocorrelation of a weak signal. This could prevent the acquisition of weak signals in the presence of strong interfering signals.

The acquisition process produces approximate Doppler shift and code delay. However, the accuracy of the estimation is not suitable to initialize the tracking modules. Therefore, fine acquisition is needed before activating the tracking modules.

Tracking of very weak signals also requires very long coherent and incoherent integrations. Long integration reduces the noise variance; thus, it increases the mean time to lose lock. It is essential to identify the bit edge position to allow the coherent integration to have a length of at least one data bit interval. Detecting the data of the navigation message is required to increase the coherent integration beyond one data bit length. Decoding the navigation message is required to calculate the navigation solution. Low-power signals have high BER, which does not allow for such decoding.

The limited resources of the wireless devices introduce another problem to the implementation of GPS receiver algorithms for weak signals. It is essential for such devices to have low cost and size, and they usually have small processors, small memory, and limited battery life. Furthermore, the algorithms should not require the collection of a large amount of data. Thus, any practical algorithm needs to be carefully designed to fit the requirements of such devices.

1.14.2 Overview of the Developed Algorithms

The problems associated with the design of GPS receiver algorithms for weak signals are addressed in the algorithms introduced in this book. Fifteen algorithms are developed; these algorithms are as follows:

- *Two weak-signal acquisition algorithms using coherent integration that is a multiple of one data bit interval.* They differ in the method used to calculate the coherent integration. One algorithm uses circular correlation, and the other uses DBZP. The most likely data bit combination is estimated in each coherent integration interval and used to remove the effect of the data bit signs. The increased processing due to the increased coherent integration length is circumvented by starting the acquisition with a small coherent integration, eliminating the unlikely Doppler bins, and increasing the coherent integration length. The DBZP is modified to mitigate the limitations on the lengths of both the coherent and incoherent integrations. In addition, two other algorithms are developed; one algorithm acquires weak signals in the presence of strong interfering signals, and one algorithm acquires weak signals under high dynamic conditions. The probabilities of false alarm and detection are derived taking into consideration all of the approaches used in these algorithms. These algorithms are presented in Chapter 3.
- *A fine acquisition algorithm based on the VA.* This algorithm works iteratively to reduce the processing requirement. A high dynamic fine acquisition is also developed to utilize the relationship between the unknown Doppler shift and Doppler rate errors; thus, it avoids performing

a search on all of the possible combinations of these errors. These algorithms are presented in Chapter 4.

- *A bit synchronization and data-detection algorithm based on both the VA and the EKF.* EKFs are used in each of the VA survivor paths to estimate and remove the carrier errors from the correlator complex output before calculating the VA path weights. This aims to achieve high edge detection rate (EDR) and optimal BER. This algorithm is presented in Chapter 4.
- *Code- and carrier-tracking algorithms based on the EKF.* Several EKF designs are introduced, including first-order, second-order, and square root EKFs. The algorithms use an integration multiple of one data bit interval. The integration length changes adaptively based on the correctly decoded data. The code-tracking module can change the early/late code-delay separation with time. More than one code-delay separation can be used at the same time. In addition, methods are introduced to detect large code- and carrier-tracking errors, and then these errors are corrected by, respectively, acquisitionlike and reinitialization algorithms. This is to avoid losing lock completely and having to reacquire the signal. Also, an algorithm is developed that has the ability to track signals in situations where the receiver dynamics encounter large, sudden changes. These algorithms are presented in Chapter 5.
- *Algorithms that decode the navigation message of signals with high BERs that do not allow for message decoding.* The navigation message structure is utilized to enable message decoding. First, the preamble and the subframe IDs are identified, and then the rest of the message is decoded. This enables the navigation solution to be calculated for weak signals. These algorithms are presented in Chapter 5.

None of the algorithms requires any assisting information. The processing and memory requirements are considered in their design.

References

[1] Parkinson, B., and J. Spilker, *Global Positioning System: Theory and Applications*, Washington D.C.: AIAA, 1996.

[2] Misra, P., and P. Enge, *Global Positioning System: Signals, Measurements, and Performance*, Lincoln, MA: Ganga-Jumuna Press, 2001.

[3] Kaplan, E., *Understanding GPS: Principles and Applications*, Norwood, MA: Artech House, 1996.

[4] NAVSTAR GPS Space Segment, Navigation User Interface Control Document (ICD-GPS-200), 1993–2000.

[5] Tsui, J. B. Y., *Global Positioning System Receivers: A Software Approach*, New York: John Wiley & Sons, 2000.

[6] Van Nee, D. J. R., and A. J. R. M. Coenen, "New Fast GPS Code-Acquisition Technique Using FFT." *IEEE Electronics Letters*, Vol. 27, No. 2, January 17, 1991.

[7] Lin, D. M., and J. B. Y. Tsui, "Acquisition Schemes for Software GPS Receiver." *Proc. ION GPS*, Nashville, TN, September 15–18, 1998, pp. 317–325.

[8] Jones, J. J., "Recovery of a Timing Reference by Square-Law Processing of a Random Pulse Train in Noise," Applied Research Laboratory, Sylvania Electronic Systems, Research Report 488, Dec. 1965.

[9] Lin, D. M., and J. B. Y. Tsui, "Comparison of Acquisition Methods for Software GPS Receiver," *Proc. ION GPS*, Salt Lake City, UT, September 19–22, 2000, pp. 2385–2390.

[10] Lin, D. M., J. B. Y. Tsui, and D. Howell, "Direct P(Y)-Code Acquisition Algorithm for Software GPS Receivers," *Proc. ION GPS*, Nashville, TN, September 14–17, 1999, pp. 363–367.

[11] Lin, D. M., and J. B. Y. Tsui, "An Efficient Weak Signal Acquisition Algorithm for a Software GPS Receiver," *Proc. ION GPS*, Salt Lake City, UT, September 11–14, 2001, pp. 115–119.

[12] Starzyk, J. A., and Z. Zhu, "Averaging Correlation for C/A Code Acquisition and Tracking in Frequency Domain," *Proc. 44th IEEE 2001 Midwest Symposium on Circuits and Systems MWSCAS 2001*, Vol. 2, August 14–17, 2001, Dayton, OH, pp. 905–908.

[13] Lin, D. M., and J. B. Y. Tsui, "A Software GPS Receiver for Weak Signals," *Proc. IEEE MTT-S Digest*, 2001, Phoenix, AZ, May 20–25, 2001, pp. 2139–2142.

[14] Psiaki, M. L., "Block Acquisition of Weak GPS Signals in a Software Receiver," *Proc. ION GPS*, Salt Lake City, September 11–14, 2001, 2838–2850.

[15] Akos, D. M., et al., "Low Power Global Navigation Satellite System (GNSS) Signal Detection and Processing," *Proc. ION GPS*, Salt Lake City, UT, September 19–22, 2000, pp. 784–791.

[16] Chansarkar, M. M., and L. Garin, "Acquisition of GPS Signals at Very Low Signal to Noise Ratio," *Proc. ION NTM 2000*, Anaheim, CA, January 26–28, 2000, pp. 731–737.

[17] Mengali, U., and A. N. D'Andrea, *Synchronization Techniques for Digital Receivers*, New York: Plenum Press, 1997.

[18] Lindsey, W. C., and M. K. Simon, *Telecommunication Systems Engineering*, Upper Saddle River, NJ: Prentice Hall, 1973.

[19] Million, S., and S. Hinedi, "Effects of Symbol Transition Density on the Performance of the Data Transition Tracking Loop at Low Signal-to-Noise Ratios," *Proc. IEEE International Conference on Communications ICC*, Vol. 2, Seattle, WA, 1995, pp. 1036–1040.

[20] Pomalaza-Raez, C. A., "Analysis of an All-Digital Data-Transition Tracking Loop," *IEEE Transactions on Aerospace and Electronic Systems*, Vol. 28, No. 4, October 1992, pp. 1119–1127.

[21] Vesma, J., et al., "Block-Based Feedforward Maximum Likelihood Symbol Timing Recovery Technique." Telecommunications Laboratory, Tampere University of Technology, P. O.

Box 553, FIN-33101 Tampere, Finland. Nokia Research Center, P.O. Box 100, FIN-33721 Tampere, Finland.

[22] Oerder, M., and H. Meyr, "Digital Filter and Square Timing Recovery," *IEEE Transactions on Communications*, Vol. 36, No. 5, May 1988, pp. 605–612.

[23] Kokkonen, M., and S. Pietila, "A New Bit Synchronization Method for a GPS Receiver," *Proc. IEEE Positioning, Location, and Navigation Symposium (PLANS)*, Palm Springs, CA, April 2002, pp. 85–90.

[24] Hill, J., and W. Michalson, "Real Time Verification of Bit-Cell Alignment for C/A Code Only Receivers," *Proc. ION Annual Meeting*, Albuquerque, NM, June 2001, pp. 463–468.

[25] Rezeanu, S. C., R. E. Ziemer, and M. A. Wickert, "Joint Maximum-Likelihood Parameter Estimation for Burst DS Spread-Spectrum Transmission," *IEEE Transactions on Communications*, Vol. 45, No. 2, February 1997, pp. 227–238.

[26] Rezeanu, S. C., and R. E. Ziemer, "Fast-Acquisition Algorithm for Joint Maximum-Likelihood Data and Bit Synchronization Epoch Estimation in Burst Direct-Sequence Spread-Spectrum Transmission," *IEEE Communications Letters*, Vol. 2, No. 2, February 1998, pp. 33–35.

[27] Zhengdi, Q., "Self-Aiding in GPS Signal Tracking," *Proc. ION GPS*, Portland, OR, September 24–27, 2002, pp. 506–509.

[28] Legrand, F., and C. Macabiau, "Improvement of the Tracking Capabilities of GPS Receivers in Weak Signal Environment and without Data Demodulation," *Proc. 19th AIAA ICSSC*, Toulouse, France, April 17–20, 2001.

[29] Akopian, D., and J. Syrjarinne, "A Network Aided Iterated LS Method for GPS Positioning and Time Recovery without Navigation Message Decoding," In *Proc. IEEE Positioning, Location, and Navigation Symposium (PLANS)*, Palm Springs, CA, April 2002, pp. 77–84.

[30] Sirola, N., and J. Syrjarinne, "GPS Position Can Be Computed without the Navigation Message," *Proc. ION GPS*, Portland, OR, September 24–27, 2002, pp. 2741–2744.

[31] Psiaki, M. L., and H. Jung, "Extended Kalman Filter Methods for Tracking Weak GPS Signals," *Proc. ION GPS*, Portland, OR, September 24–27, 2002, pp. 2539–2553.

[32] Lin, D. M., and J. B. Y. Tsui, "A Weak Signal Tracking Technique for a Stand-Alone Software GPS Receiver," *Proc. ION GPS*, Portland, OR, September 24–27, 2002, pp. 2534–2538.

[33] Psiaki, M. L., "Smoother-Based GPS Signal Tracking in a Software Receiver," *Proc. ION GPS*, Salt Lake City, UT, September 11–14, 2001, pp. 2900–2913.

[34] Tsui, J. B. Y., M. H. Stockmaster, and D. M. Akos, "Block Adjustment of Synchronizing Signal (BASS) for Global Positioning System (GPS) Receiver Signal Processing," *Proc. ION GPS*, 1997, Kansas City, MD, Sept. 16–19, 1997, pp. 637–643.

[35] Gustafson, D. E., "GPS Signal Tracking Using Maximum-Likelihood Parameter Estimation," *Journal of Navigation*, Vol. 45, No. 4, 1999, pp. 287–296.

[36] Zhodzishaky, M., et al., "Co-Op Tracking for Carrier Phase," *Proc. ION GPS*, 1998, Nashville, TN, Sept. 15–18, 1998, pp. 653–664.

[37] Hong, J. S., et al., "GPS Signal Processing Algorithm for Software GPS Receiver," *Proc. ION GPS*, 2000, Salt Lake City, UT, Sept. 19–22, 2000, pp. 2338–2345.

[38] Schamus, J. J., and J. B. Y. Tsui, "Acquisition to Tracking and Coasting for Software GPS Receiver," *Proc. ION GPS*, 1999, Nashville, TN, Sept. 14–17, 1999, pp. 325–328.

[39] Galileo Joint Undertaking, "Galileo Signal in Space ICD (SIS ICD)," *3GPP TSG GERAN2* meeting, May 23–27, 2005, Quebec, Canada, Reference Idoc G2-050266.

[40] Hein, G. W., et al., "Status of Galileo Frequency and Signal Design," *Proc. ION GPS*, Portland, OR, September 24–27, 2002, pp. 266–277.

[41] Richton G. M., and R. E. Djuknic, "Geolocation and Assisted GPS," *IEEE Computer*, Vol. 34, No. 2, February 2001, pp. 123–125.

[42] Bajaj, R., S. L. Ranaweera, and D. P. Argrawal, "GPS: Location-Tracking Technology," *IEEE Computer*, Vol. 35, No. 4, March 2002, pp. 92–94.

[43] Zhao, Y., "Mobile Phone Location Determination and Its Impact on Intelligent Transportation Systems," *IEEE Trans. Intelligent Transportation Systems*, Vol. 1, No. 1, March 2000, pp. 55–64.

[44] Van Diggelen, F., "Indoor GPS Theory and Implementation," *Proc. IEEE Positioning, Location, and Navigation Symposium (PLANS)*, 2002, Palm Springs, CA, April 15–18, 2002, pp. 240–247.

[45] Van Diggelen, F., and C. Abraham, "Method and Apparatus for Enhancing a Global Positioning System with Terrain Model," U.S. Patent No. 6,429,814, August 6, 2002.

[46] Fuchs, V. L., C. Abraham, and F. Van Diggelen, "Method and Apparatus for Locating and Providing Services to Mobile Devices," U.S. Patent No. 6,453,237, September 17, 2002.

[47] LaMance, J. W., C. Abraham, and F. Van Diggelen, "Method and Apparatus for Generating and Distributing Satellite Tracking Information," U.S. Patent No. 6,542,820, April 1, 2003.

[48] Van Diggelen, F., "Method and Apparatus for Locating Mobile Receivers Using a Wide Area Reference Network for Propagating Ephemeris," U.S. Patent No. 6,587,789, July 1, 2003.

[49] Abraham, C., and V. L. Fuchs, "Method and Apparatus for Computing Signal Correlation at Multiple Resolutions," U.S. Patent No. 6,704,348, March 9, 2004.

[50] Van Diggelen, F., "Method and Apparatus for Processing of Satellite Signals without Time of Day Information," U.S. Patent No. 6,734,821, May 11, 2004.

[51] Abraham, C., and F. Van Diggelen, "Method and Apparatus for Performing Signal Correlation Using Historical Correlation Data," U.S. Patent No. 6,819,707, November 16, 2004.

[52] Krasner, N. F., Method and Apparatus for Acquiring Satellite Positioning System Signals," U.S. Patent No. 6,133,874, October 17, 2000.

[53] Krasner, N. F., "GPS Receiver Utilizing a Communication Link," U.S. Patent No. 6,421,002, July 16, 2002.

[54] Krasner, N. F., "Highly Parallel GPS Correlator System and Method," U.S. Patent No. 6,208,291, March 27, 2001.

[55] Krasner, N. F., "GPS Receivers and Garments Containing GPS Receivers and Methods for Using These GPS Receivers," U.S. Patent No. 6,259,399, July 10, 2001.

[56] Krasner, N. F., "GPS Receiver and Method for Processing GPS Signals," U.S. Patent No. 6,725,159, April 20, 2004.

[57] Sheynblat, L., "Methods and Apparatuses for Using Assistance Data Relating to Satellite Position Systems," U.S. Patent No. 6,720,915, April 13, 2004.

[58] Krasner, N. F., "Method and Apparatus for Signal Processing in a Satellite Positioning System," U.S. Patent No. 6,816,710, November 9, 2004.

[59] Chong, C., and S. P. Kumar, "Sensor Networks: Evolution, Opportunities, and Challenges," *Proceedings of the IEEE,* Vol. 91, No. 8, August 2003, pp. 1247–1256.

[60] Bulusu, N., J. Heidemann, and D. Estrin, "GPS-Less Low-Cost Outdoor Localization for Very Small Devices," *IEEE Personal Communications,* Vol. 7, No. 5, October 2000, pp. 28–34.

[61] Ergen, M., et al., "Application of GPS to Mobile IP and Routing in Wireless Networks," *Proc. IEEE 56th Vehicular Technology Conference,* Vol. 2, September 2002, pp. 1115–1119.

[62] Cormen, T., C. Leisersen, and R. Rivest, *Introduction to Algorithms,* Boston: MIT Press, 1990.

[63] Macchi, O., and L. Scharf, "A Dynamic Programming Algorithm for Phase Estimation and Data Decoding on Random Phase Channels," *IEEE Trans Information Theory,* Vol. 27, No. 5, 1981, pp. 581–595.

[64] Vanelli-Coralli, A., et al., "A Performance Review of PSP for Joint Phase/Frequency and Data Estimation in Future Broadband Satellite Networks," *IEEE Trans Selected Areas in Communications,* Vol. 19, December 2001, pp. 2298–2309.

[65] Esteves, E. S., and R. Sampaio-Neto, "A Per-Survivor Phase Acquisition and Tracking Algorithm for Detection of TCM Signals with Phase Jitter and Frequency Error," *IEEE Trans Communications,* Vol. 45, November 1997, pp. 1381–1384.

[66] Meyr, H., M. Moeneclaey, and S. Fechtel, *Digital Communication Receivers: Synchronization, Channel Estimation, and Signal Processing,* New York: Wiley Interscience Publications, 1998.

[67] Kalman, R. E., A New Approach to Linear Filtering and Prediction Problems, *Transactions of the ASME, Journal of Basic Engineering,* Vol. 82(Series D), 1960, pp. 35–45.

[68] Maybeck, P. S., *Stochastic Models, Estimation, and Control,* Vol. 1, Burlington, MA: Academic Press, 1979.

[69] Zarchan, P., and H. Musoff. *Fundamentals of Kalman Filtering: A Practical Approach,* Washington, D.C.: AIAA, 2000.

2

Signal Models

2.1 Introduction

This chapter discusses the effect of the Doppler shift on code length and derives all the signal models that are used in this book. Section 2.2 presents a model for the received signal. Section 2.3 presents an overview of the Doppler effect on the code length. Section 2.4 presents a model for the Doppler-compensated replica local signals. Then, Sections 2.5 to 2.7 present the correlation output of the local signal generator at different stages of the GPS signal processing. The model in Section 2.5 is used for signal acquisition. The model in Section 2.6 is used for fine acquisition and bit synchronization; it assumes that an approximate Doppler shift, Doppler rate, and code delay are available. The model in Section 2.7 is used for tracking; it assumes that the estimated parameters are continuously updated. A different model for the correlation output is used at each stage of receiver processing to relate the output of each stage to the input of the following stage. Section 2.8 presents a model for clock noise disturbances.

2.2 Received Signals

The sampled received IF signal, with a sampling rate of f_s, is modeled as

$$r(t_k) = \sum_{sat=1}^{N_{sat}} \Big\{ A_{sat}\, d_{sat}\big(t_k, \tau_{sat}, fd_{sat}\big)\, C_{sat}\big(t_k, \tau_{sat}, fd_{sat}\big) \cos\big(\theta_{n_k} + \theta_{sat,0} + 2\pi\, \big(f_{IF} + fd_{sat,0}\big)\, t_k + \pi\, \alpha_{sat}\, t_k^2\big)\Big\} + n(t_k) \quad (2.1)$$

where N_{sat} is the number of visible satellites. The subscript *sat* is used to distinguish between different satellites; this subscript will be dropped in the rest of the book. t_k is the sampling time. A is the signal amplitude; it is normalized to drive the noise variance, σ_n^2, to 1 such that $A = \sqrt{2a\ S/N_0\ T_s}$; $T_s = 1/f_s$; a is a constant. $S/N_0 = 10^{(C/N_0)/10}$; C/N_0 is the carrier-to-noise ratio in dB-Hz units. d is the navigation data. C is the received PRN code, which has a length of 1,023 chips and a chipping rate of 1.023 MHz. θ_0 is the initial phase. θ_{n_k} is the phase noise at time t_k; it is composed of the total phase and frequency clock disturbances. f_{IF} is the IF carrier frequency. f_{d_0} is the Doppler shift at $t_k = 0$. α is the Doppler rate. n is a white Gaussian noise (WGN) with zero mean and unit variance.

The Doppler shift changes the rate of the PRN code or, equivalently, the length of the PRN code. This effect must be taken into consideration; therefore, the sampling time of the received PRN code and the data is expressed as follows:

$$t_{k,\tau,f_d} = (t_k - \tau)\left(1 + \frac{f_{d_0} + \alpha\ t_k/2}{f_{L1}}\right) \tag{2.2}$$

where τ is the code delay, and f_{L1} is the L1 carrier frequency. The Doppler effect on the code rate is explained in the next section.

2.3 Doppler Effect on the Code Rate

In [1] the Doppler shift is derived on the L1 frequency as f_d and on the C/A code as f_{dc}:

$$f_d = f_{L1}\frac{-v_r}{c} \tag{2.3}$$

$$f_{dc} = f_{code}\frac{v_r}{c} \tag{2.4}$$

where v_r is the range rate between the satellite and the receiver, c is the speed of light, f_{L1} is the L1 carrier frequency, and f_{code} is the C/A code frequency. From (2.3) and (2.4), the Doppler shift on the code is

$$f_{dc} = -f_{code}\frac{f_d}{f_{L1}} \tag{2.5}$$

The length of one C/A code is $1/f_{code}$. The Doppler effect will cause the C/A code to shift one cycle every $1/f_{dc}$ sec. So, the Doppler effect on the code length causes a shift of f_{shift} Hz in T sec., which can be calculated from

$$f_{shift} = -\frac{1}{T}\frac{1}{f_{dc}} f_{code} \tag{2.6}$$

or, equivalently,

$$f_{shift} = -\frac{1}{T}\frac{f_{L1}}{f_d} \tag{2.7}$$

This effect is equivalent to a change in the code delay. The amount of the change in the code delay $\Delta\tau$ in T sec. can be calculated from

$$\Delta\tau = -T\frac{f_d}{f_{L1}} \tag{2.8}$$

A positive Doppler shift causes the code length to shrink, while a negative Doppler shift causes the code length to expand. This effect can be expressed as [2]

$$T_{c_d} = T_c\frac{f_{L1}}{f_{L1} + f_d} \tag{2.9}$$

where T_{c_d} is the new code length, and T_c is the original code length. If the code delay is τ, and the code is sampled at time instances t_k, then the actual location of the samples, defined as t_{k,τ,f_d}, in the code can be calculated from

$$t_{k,\tau,f_d} = (t_k - \tau)\left(1 + \frac{f_d}{f_{L1}}\right) \tag{2.10}$$

If the Doppler shift changes at a rate of α, then (2.10) is modified as

$$t_{k,\tau,f_d} = (t_k - \tau)\left(1 + \frac{f_{d_0} + \alpha t_k/2}{f_{L1}}\right) \tag{2.11}$$

where f_{d_0} is the Doppler shift at $t_k = 0$. If f_{d_i} defines the Doppler shift at the start of the ith T_c interval, then the Doppler-compensated code length over the ith interval is

$$T_{c_i} = T_c\frac{f_{L1}}{f_{L1} + f_{d_i} + \alpha T_c/2} \tag{2.12}$$

This means that if $\alpha \neq 0$, then the received code length will change as α changes.

2.4 Local Signals

The receiver generates a local replica PRN code for a satellite and uses it within the acquisition and tracking modules. The in-phase, I_L, and quad-phase, Q_L, Doppler-compensated local replica code can be expressed by the following model:

$$I_L(t_k, \tau_v, f_{d_u}) = C_L(t_{k,\tau_v, f_{d_u}}) \cos(\theta_{L,0} + 2\pi\,(f_{IF} + f_{d_{u,0}})\, t_k + \pi\, \alpha_L\, t_k^2) \tag{2.13}$$

$$Q_L(t_k, \tau_v, f_{d_u}) = C_L(t_{k,\tau_v, f_{d_u}}) \sin(\theta_{L,0} + 2\pi\,(f_{IF} + f_{d_{u,0}})\, t_k + \pi \alpha_L t_k^2) \tag{2.14}$$

where C_L is the local PRN code, and τ_v, $f_{d_{u,0}}$, α_L, and $\theta_{L,0}$ are the locally generated code delay, Doppler shift at $t_k = 0$, Doppler rate, and phase at $t_k = 0$, respectively. C_L can be generated by accounting for the Doppler effect on the code length, so

$$t_{k,\tau_v, f_{d_u}} = (t_k - \tau_v)\left(1 + \frac{f_{d_{u,0}} + \alpha_L\, t_k/2}{f_{L1}}\right) \tag{2.15}$$

If the Doppler effect on the code length is not considered, then $t_{k,\tau_v, f_{d_u}}$ is replaced by t_{k,τ_v} such that

$$t_{k,\tau_v} = (t_k - \tau_v) \tag{2.16}$$

2.5 Signal Model for Acquisition

The acquisition process assumes that both θ_{L_0} and α_L are equal to 0. The receiver performs an integrate-and-dump process using the received signal and the local signals. This process is equivalent to

$$I_i(\tau_v, f_{d_u}) = \sqrt{\frac{1}{N_{T_i}}} \sum_{k=si}^{si+N_{T_i}-1} I_L(t_k, \tau_v, f_{d_u})\, r(t_k) \tag{2.17}$$

$$Q_i(\tau_v, f_{d_u}) = \sqrt{\frac{1}{N_{T_i}}} \sum_{k=si}^{si+N_{T_i}-1} Q_L(t_k, \tau_v, f_{d_u})\, r(t_k) \tag{2.18}$$

where N_{T_i} and si are, respectively, the number of the samples and the index of the first sample in the ith interval. The factor $\sqrt{1/N_{T_i}}$ is used to keep the noise variance equal to one. If the received and the local codes are the same, then the output will have the following model:

$$I_i\left(\tau_v, f_{d_u}\right) = A\, d_i\, R(\tau - \tau_v)\ \mathrm{sinc}\left(\left(f_{d_i} + \alpha \frac{T_i}{2} - f_{d_u}\right) T_i\right)\ \cos(\theta_{e_i}) + n_{I_i} \tag{2.19}$$

$$Q_i\left(\tau_v, f_{d_u}\right) = A\, d_i\, R(\tau - \tau_v)\ \mathrm{sinc}\left(\left(f_{d_i} + \alpha \frac{T_i}{2} - f_{d_u}\right) T_i\right)\ \sin(\theta_{e_i}) + n_{Q_i} \tag{2.20}$$

where T_i is the length of the ith interval, f_{d_i} is the received Doppler shift at the beginning of the ith interval, θ_{e_i} is the average phase error over the ith interval, and $R(\tau - \tau_v)$ is the autocorrelation function. If $|\tau - \tau_v| < T_{chip}$, then $R(\tau - \tau_v)$ is proportional to $\left(1 - |\tau - \tau_v|/T_{chip}\right)$, where T_{chip} is the length of one code chip; otherwise $R(\tau - \tau_v)$ is equal to zero. n_I and n_Q are independent WGN noise with zero mean unit variance.

2.6 Signal Model for Fine Acquisition

After an acquisition process, an approximate Doppler shift $\hat{f}_d$, Doppler rate $\hat{\alpha}$, phase $\hat{\theta}$, and code delay $\hat{\tau}$ are obtained. They are used as inputs to the local signal generator, which in turn generates signals for the fine acquisition module. The model for the in-phase and quad-phase local signals can be derived from (2.13) and (2.14) as

$$I_L(t_k) = C_L\left(t_{k,\hat{\tau},\hat{f}_d}\right)\ \cos\left(\hat{\theta}_0 + 2\pi\left(f_{IF} + \hat{f}_{d_0}\right) t_k + \pi\,\hat{\alpha}\, t_k^2\right) \tag{2.21}$$

$$Q_L(t_k) = C_L\left(t_{k,\hat{\tau},\hat{f}_d}\right)\ \sin\left(\hat{\theta}_0 + 2\pi\left(f_{IF} + \hat{f}_{d_0}\right) t_k + \pi\,\hat{\alpha}\, t_k^2\right) \tag{2.22}$$

where $\hat{\theta}_0$ and $\hat{f}_{d_0}$ are the estimated phase and Doppler shift at $t_k = 0$. The time instance $t_{k,\hat{\tau}}$ is modeled as follows:

$$t_{k,\hat{\tau},\hat{f}_d} = (t_k - \hat{\tau})\left(1 + \frac{\hat{f}_{d_0} + \hat{\alpha}\, t_k/2}{f_{L1}}\right) \tag{2.23}$$

From (2.2) and (2.23), the following can be obtained:

$$\begin{aligned} t_{k,\tau,f_d} - t_{k,\hat{\tau},\hat{f}_d} &= (t_k - \tau,\ f_d) \left(1 + \frac{f_{d_0} + \alpha\ t_k/2}{f_{L1}}\right) - (t_k - \hat{\tau}) \\ &\quad \times \left(1 + \frac{\hat{f}_{d_0} + \hat{\alpha}\ t_k/2}{f_{L1}}\right) \\ &= t_k \left(\frac{f_{d_0} + \alpha\ t_k/2}{f_{L1}} - \frac{\hat{f}_{d_0} + \hat{\alpha}\ t_k/2}{f_{L1}}\right) \\ &\quad -\tau \left(1 + \frac{f_{d_0} + \alpha\ t_k/2}{f_{L1}}\right) + \hat{\tau} \left(1 + \frac{\hat{f}_{d_0} + \hat{\alpha}\ t_k/2}{f_{L1}}\right) \end{aligned} \tag{2.24}$$

Let

$$\tau_e = \tau - \hat{\tau} \tag{2.25}$$

$$f_{e_0} = f_{d_0} - \hat{f}_{d_0} \tag{2.26}$$

$$\alpha_e = \alpha - \hat{\alpha} \tag{2.27}$$

So, (2.24) becomes

$$\begin{aligned} t_{k,\tau,f_d} - t_{k,\hat{\tau},\hat{f}_d} &= t_k \left(\frac{f_{e_0} + \alpha_e\ t_k/2}{f_{L1}}\right) \\ &\quad -\hat{\tau} \left(\frac{f_{e_0} + \alpha_e\ t_k/2}{f_{L1}}\right) - \tau_e \left(1 + \frac{f_{d_0} + \alpha\ t_k/2}{f_{L1}}\right) \\ &= (t_k - \hat{\tau}) \left(\frac{f_{e_0} + \alpha_e\ t_k/2}{f_{L1}}\right) - \tau_e \left(1 + \frac{f_{d_0} + \alpha\ t_k/2}{f_{L1}}\right) \end{aligned} \tag{2.28}$$

Thus, the Doppler shift and Doppler rate estimation errors will cause a change in the code-delay error with time.

Define τ_{e_i} as the code-delay error at the ith interval and θ_{e_i} as the average phase error over the ith interval. Also, let

$$f_{e_i} = f_{d_i} - \hat{f}_{d_i}. \tag{2.29}$$

where $\hat{f}_{d_i}$ is the estimated Doppler shift at the start of the ith interval. This can be found from

$$\hat{f}_{d_i} = \hat{f}_{d_0} + i\ T_i \hat{\alpha} \tag{2.30}$$

The correlated signal models in (2.19) and (2.20) are modified as

$$I_i = A\, d_i\, R(\tau_{e_i})\ \mathrm{sinc}\left(\left(f_{e_i} + \alpha_e \frac{T_i}{2}\right) T_i\right)\ \cos(\theta_{e_i}) + n_{I_i} \tag{2.31}$$

$$Q_i = A\, d_i\, R(\tau_{e_i})\ \mathrm{sinc}\left(\left(f_{e_i} + \alpha_e \frac{T_i}{2}\right) T_i\right)\ \sin(\theta_{e_i}) + n_{Q_i} \tag{2.32}$$

2.7 Signal Model for Tracking

The code- and carrier-tracking modules make use of the prompt, early, and late signals. These signals are generated with integration length equal to T_i, where the length of the T_i interval can be equal to either one code length or a multiple of one code length. The start of each interval is the estimated start of the first code in that interval. Since the first sample in each interval does not have to be located exactly on the start of a code, the first sample in each interval represents the estimated first sample of the first code in that interval. Define the time of this first sample in the ith interval as t_{i_0}. The difference between the models of the received and local signals is that the time of the received sample model is calculated relative to the start of the tracking, while the time of the local sample model is calculated relative to the start of an interval. Define t_{s_i} as the difference between the time of the first sample in the ith interval and the time of the start of the first code in that interval. So, if f_s is the sampling rate, then $0 \le t_{s_i} < 1/f_s$. Figure 2.1 shows the timing relationship between the received code and the locally generated prompt, late, and early codes.

The local samples in each interval are generated using the updated estimates of the code delay, phase, Doppler shift, and Doppler rate. Let $T_s = 1/f_s$.

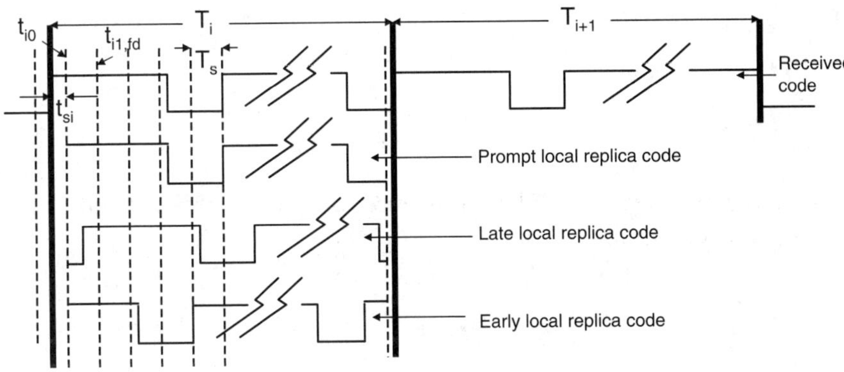

Figure 2.1 Timing relationship between the received code and the locally generated prompt, late, and early replica code versions.

The received sample model for one satellite over the ith interval can be derived from (2.1) as

$$r(t_{i_0}+lT_s) = A\,d(t_{il,f_d})\,C(t_{il,f_d})\cos\left(\theta_{n_{t_{il}}}+\theta_0+2\pi(f_{IF}+f_{d_0})(t_{i_0}+lT_s)\right.$$
$$\left.+\pi\,\alpha(t_{i_0}+lT_s)^2\right)+n(t_{i_0}+lT_s) \qquad (2.33)$$

where l is the index of the lth sample within an interval $l=0,\ldots,N_{T_i}-1$. N_{T_i} is the number of samples in the ith interval. f_{d_0} is the received Doppler shift at the start of the tracking. $\theta_{n_{t_{il}}}$ is the accumulated phase noise, and

$$t_{il,f_d} = (t_{i_0}+lT_s-(\tau_i-\hat{\tau}_i))\left(1+\frac{f_{d_0}+\alpha(t_{i_0}+lT_s)/2}{f_{L1}}\right) \qquad (2.34)$$

where $\hat{\tau}_i$ is the estimated code delay at the ith interval. The model of the in-phase and quad-phase local samples over the ith interval are

$$\begin{aligned} I_L(t_{i_0}+lT_s+\delta) &\equiv I_L(t_{s_i}+lT_s+\delta) \\ &= C_L(t_{l,\delta,\hat{f}_d})\cos\left(\hat{\theta}_i+2\pi(f_{IF}+\hat{f}_{d_i})\right. \\ &\quad\left.\times(t_{s_i}+lT_s+\delta)+\pi\,\hat{\alpha}_i(t_{s_i}+lT_s+\delta)^2\right) \end{aligned} \qquad (2.35)$$

$$\begin{aligned} Q_L(t_{i_0}+lT_s+\delta) &\equiv Q_L(t_{s_i}+lT_s+\delta) \\ &= C_L(t_{l,\delta,\hat{f}_d})\sin\left(\hat{\theta}_i+2\pi(f_{IF}+\hat{f}_{d_i})\right. \\ &\quad\left.\times(t_{s_i}+lT_s+\delta)+\pi\,\hat{\alpha}_i(t_{s_i}+lT_s+\delta)^2\right) \end{aligned} \qquad (2.36)$$

where δ is the amount of shift induced in the replica code to generate the prompt, early, and late signals; $\hat{\theta}_i$ and $\hat{f}_{d_i}$ are the estimated phase and Doppler shift at the start of the ith interval; $\hat{\alpha}_i$ is the estimated Doppler rate at the ith interval; and

$$t_{l,\delta,\hat{f}_d} = (t_{s_i}+lT_s+\delta)\left(1+\frac{\hat{f}_{d_i}+\hat{\alpha}_i(t_{s_i}+lT_s+\delta)/2}{f_{L1}}\right) \qquad (2.37)$$

The local samples are correlated with the received ones as in (2.17) and (2.18). The accumulated signals are

$$I_i = A\,d_i\,R(\tau_{e_i})\,\mathrm{sinc}\left(\left(f_{e_i}+\alpha_{e_i}\frac{T_i}{2}\right)T_i\right)\cos(\theta_{e_i})+n_{I_i} \qquad (2.38)$$

$$Q_i = A\,d_i\,R(\tau_{e_i})\,\mathrm{sinc}\left(\left(f_{e_i}+\alpha_{e_i}\frac{T_i}{2}\right)T_i\right)\sin(\theta_{e_i})+n_{Q_i} \qquad (2.39)$$

where τ_{e_i} and α_{e_i} are the estimation errors over the ith interval for the code delay and Doppler rate, respectively; θ_{e_i} is the average phase error over the ith interval; and f_{e_i} is the Doppler shift estimation error at the start of the ith interval. The sinc function causes attenuation in the I_i and Q_i accumulated signals. Figure 2.2 shows the sinc function versus the frequency error. As can be seen, the sinc function is zero when $f_e = N/T_i$, where $N = 1, 2, \ldots,$ but it attenuates after the first lobe.

2.8 Clock Noise

Clock noise disturbances are modeled as normal random walks. Discussion of clock noise modeling can be found in [3]. Discussion of GPS clock noise can be found in [4, 5]. Two independent sequences W_θ and W_{f_d} are generated for the phase and frequency disturbances, respectively. Those sequences have zero mean and variances of σ_θ^2 and σ_{fd}^2. The variances have values based on the clock type [e.g., temperature-compensated crystal oscillator (TCXO) or ovenized crystal oscillator (OXO)]. The clock noises can be expressed by the phase and frequency random walk intensities S_f (in seconds) and S_g (in 1/seconds), respectively. The phase and the Doppler shift of the received IF signal samples have the following models:

$$f_{d_k} = f_{d_{k-1}} + \alpha\ T_s + W_{f_d,k} \tag{2.40}$$

$$\theta_k = \theta_{k-1} + 2\pi\ f_{d_k}\ T_s + W_{\theta,k} \tag{2.41}$$

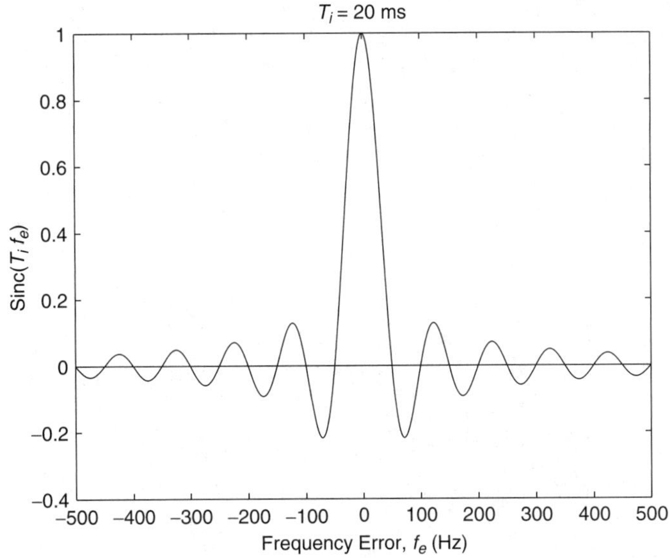

Figure 2.2 Illustration of the sinc function versus frequency error.

References

[1] Tsui, J. B. Y., *Global Positioning System Receivers: A Software Approach*, New York: John Wiley & Sons, 2000.

[2] Lin, D. M., and J. B. Y. Tsui, "A Weak Signal Tracking Technique for a Stand-Alone Software GPS Receiver," *Proc. ION GPS*, Portland, OR, September 24–27, 2002, pp. 2534–2538.

[3] Brown, R. G., and P. Y. C. Hwang, *Introduction to Random Signals and Applied Kalman Filtering*, New York: John Wiley & Sons, 1992.

[4] Parkinson, B., and J. Spilker, *Global Positioning System: Theory and Applications*, Washington, D.C.: AIAA, 1996.

[5] Kaplan, E., *Understanding GPS: Principles and Applications*, Norwood, MA: Artech House, 1996.

3

Signal Acquisition

3.1 Introduction

Two acquisition algorithms for weak GPS signals are developed in which long coherent and incoherent integrations are used, without requiring any assisting information. The coherent, or predetection, integration time (PIT) is a multiple of one data bit interval. The most likely data bit combination is estimated over each PIT and used to remove the data bits prior to including the coherent integration result in the incoherent summation. The problem of the unknown bit edge positions is handled by forming different parallel accumulations. Each accumulation starts at a different possible bit edge position. At the conclusion of the acquisition, the most likely code delay, Doppler shift, and bit edge are selected as the combination that maximizes the total accumulation.

The two algorithms differ in the method used to calculate the coherent integration. The first algorithm, circular correlation with multiple data bits (CCMDB), uses circular correlation (FFT/IFFT), while the second algorithm, modified double block zero padding (MDBZP), uses double block zero padding (DBZP). Each method suffers from some problems due to their long PIT. Those problems are addressed in the developed algorithms. Also, the CCMDB and MDBZP are modified and used to develop another two algorithms that address specific problems: the acquisition of weak signals in the presence of strong interfering signals and the acquisition of signals under environments in which the receiver is moving with high acceleration.

This chapter also provides analysis and derivation for the probability of false alarm, the probability of detection, and the calculation of the acquisition threshold. The analysis is done for the CCMDB and MDBZP algorithms. It models various errors associated with these two algorithms and accounts for the effect of any special techniques used. The analysis can be easily altered and applied to different algorithms.

The performances of the CCMDB and MDBZP are demonstrated and compared to two conventional acquisition algorithms. The first acquisition algorithm uses coherent integration of 10 ms, while the second uses coherent integration of 20 ms. In these two algorithms, the coherent integration is accumulated incoherently over long data to enable the acquisition of weak signals. The results show the ability of the CCMDB and MDBZP to acquire signals with very weak power that the two conventional algorithms fail to acquire using the same amount of total data.

This chapter is organized as follows. Sections 3.2 and 3.3 present the CCMDB and MDBZP acquisition algorithms. Section 3.4 presents an algorithm for signal acquisition in the presence of strong interfering signals. Section 3.5 presents the high dynamic acquisition algorithm. Section 3.6 provides analysis and derivation for the probabilities of false alarm and detection and the calculation of the acquisition threshold. Section 3.7 presents simulations and results for the developed algorithms. Section 3.8 presents the summary and conclusion.

3.2 Algorithm 1: Circular Correlation with Multiple Data Bits

The fundamental principles of the CCMDB are as follows. The PIT T_I is a multiple N_t of one data bit interval, T_{dms}. N_b cumulative, incoherent integrations are maintained and updated with new results after each coherent integration. Each integration starts at a possible bit edge position. After a total of L incoherent integrations, the acquisition will be concluded if the maximum total integration at a possible code delay and Doppler shift exceeds a predefined threshold, γ. T_I and T_{dms} are expressed in the units of milliseconds. All the frequencies are expressed in the units of Kilohertz in this chapter unless otherwise is specified. Figure 3.1 outlines this algorithm's steps.

Four problems associated with using a long PIT are addressed in this chapter: (1) the Doppler effect on the code length; (2) the unknown bit edge positions; (3) the unknown data bit values; and (4) increased processing and memory requirements due to the increased sample size and the number of possible Doppler shifts. The next few sections describe the solution to each of these problems.

3.2.1 Handling the Doppler Effect on the Code Length

From Section 2.3 and (2.8), it is concluded that the code will shift one sample after time T_{cs}, calculated as

$$T_{cs} = \frac{1}{f_s}\frac{f_{L1}}{f_d} \tag{3.1}$$

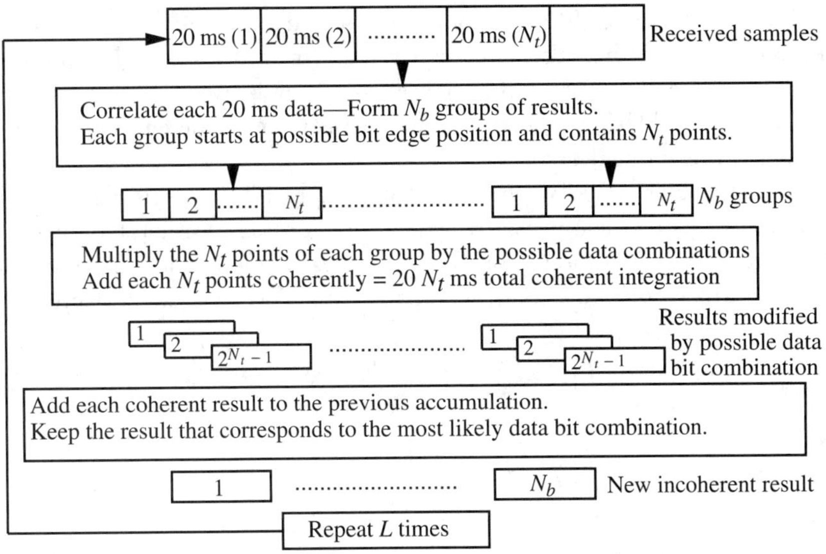

Figure 3.1 Illustration of the CCMDB acquisition algorithm's operation steps.

This means that every T_I ms, using a sampling rate of f_s KHz, the code will shift by a number of samples N_s equal to

$$N_s = f_s \; T_I \frac{f_d}{f_{L1}} \tag{3.2}$$

Figure 3.2 illustrates the problems associated with an uncompensated replica code. Assume that the upper signal is the received code and the Doppler shift is negative so that the code length increases. The lower signal is the uncompensated replica code. If the coherent integration spans the interval from A to B, and the dotted lines are the sampling times, then there are two problems. First, the

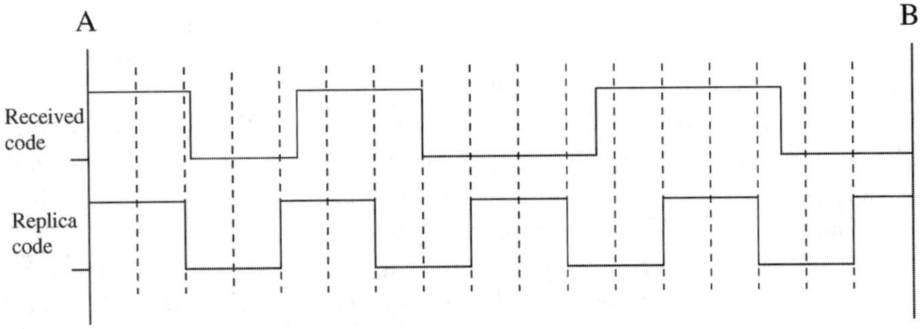

Figure 3.2 Illustration of the effect of uncompensated replica code on the integration.

samples are not taken at the same points in both codes. Second, by the end of the integration interval, the received code will have finished fewer cycles compared to the local code. This means that the relative delay between the local and the received codes will change. This will cause subsequent incoherent integration to be added at the wrong delay.

The Doppler effect on the code length is handled in this algorithm by generating a Doppler-compensated replica code for each possible Doppler shift considered by the acquisition algorithm (i.e., for each Doppler bin). Assuming the Doppler rate α is equal to zero, the compensated replica code can be expressed as

$$I_L(t_k, \tau_u, f_{d_v}) = C_L\left([t_k - \tau_u]\left[1 + \frac{f_{d_v}}{f_{L1}}\right]\right)\cos(2\pi\,[f_{IF} - f_{d_v}]\,t_k) \quad (3.3)$$

$$Q_L(t_k, \tau_u, f_{d_v}) = C_L\left([t_k - \tau_u]\left[1 + \frac{f_{d_v}}{f_{L1}}\right]\right)\sin(2\pi\,[f_{IF} - f_{d_v}]\,t_k) \quad (3.4)$$

where f_{d_v} is a possible Doppler shift, $v = 1, \ldots, N_{fd}$; and τ_u is a possible code delay, $u = 1, \ldots, N_\tau$. This compensation will guarantee that the relative delay between the received code and the local replica stays approximately the same during the acquisition process. It should be noted that T_I can be a multiple of either the compensated code length or the true code length. This will not make a difference in the performance because of the following: the replica code is compensated by the Doppler effect, so samples of the local code are taken at the correct points relative to the received code; the acquisition estimates the relative delay between the received and the local codes; and the maximum duration of the received data used in the acquisition process is a very few seconds.

3.2.2 Handling Unknown Bit Edge Positions

As a consequence of the 50-Hz data message, there are 20 possible bit edge positions. Each one is aligned with the start of a 1-ms PRN code period. If the start of the coherent integration is misaligned with respect to the start of the data bits, then there will be a loss in the integration due to the bit edge transitions.

To reduce the loss due to the misalignment, N_b arrays of incoherent integrations (accumulations) are maintained in parallel. Each of these starts at a possible bit edge position and consists of the results from $N_\tau\ N_{f_d}$ delay-Doppler cells from each possible code delay and Doppler shift. To reduce processing and memory requirements, N_b is chosen to be less than 20, with the chosen edges evenly spaced over all of the 20 possible edges.

In the worst case of misalignment, the correct bit edge will be separated by $S_{N_b} = \lfloor T_{dms}/(2N_b) \rfloor$ ms from one of the N_b edges. If there is a data transition, then an integration over T_{dms} ms will be equivalent to an integration over $T_{dms} - 2S_{N_b}$ ms without a transition. For a total coherent integration time of $T_I = N_t\, T_{dms}$, there are $N_t + 1$ data bits where the first and last data bits are incomplete. Thus, there are 2^{N_t+1} possible data bit combinations; they generate $N_t + 1$ possible equivalent integration times. These equivalent integration times can be found from $T_{I_\epsilon} = T_I - 2\epsilon\, S_{N_b}$, where $\epsilon = 0, \ldots, N_t$. Each result can occur $2[N_t!/(\epsilon!\,(N_t - \epsilon)!)]$ times in the 2^{N_t+1} possible bit combinations. Assume that the probability of each data bit combination is equal. The average equivalent integration time, in the worst case, is therefore equal to

$$T_{I_{N_b}} = \sum_{\epsilon=0}^{N_t} \frac{1}{2^{N_t}} \frac{N_t!}{\epsilon!\,(N_t - \epsilon)!}\,(T_I - 2\epsilon\, S_{N_b}) \tag{3.5}$$

Thus, the loss due to searching only N_b bit edges is $Loss_{N_b} = [100\,(T_I - T_{I_{N_b}})/T_I]\%$. For example, $Loss_{N_b}$ for $N_b = 1$, 2, and 4 equals 50%, 25%, and 10%, respectively. Thus, it can be concluded that N_b should not be chosen to be less than 4. Although choosing $4 \leq N_b < 20$ will cause some loss, it will largely reduce the processing overhead. For example, if $N_b = 4$, the reduction in the processing and memory requirements is approximately $(16/20) * 100 = 80\%$, but the worst-case integration loss is only 10%.

3.2.3 Handling Unknown Data Bits

In each step l of the acquisition, the most likely combination of N_t data bits in the interval $T_I = N_t\, T_{dms}$ is estimated. The estimated combination is then used to remove the effect of the data signs before calculating the total coherent integration at that step. This process is applied in parallel for each of the N_b possible bit edge positions. The general method for selecting the most likely data bit combination at the lth step and incorporating the new coherent integration of that step into a cumulative incoherent integration is as follows: A coherent integration of T_{dms} is performed over each of the N_t intervals in the T_I ms. This will generate N_t matrices, each of size $N_\tau \times N_{f_d}$. These N_t matrices are multiplied by the possible data bit combinations to remove the effect of the data signs. Then, the N_t resultant matrices are added together to generate the total coherent integration. Since the purpose is to add data bits with the same sign over each coherent integration interval, only 2^{N_t-1} possible data bit combinations are used. So, there will be 2^{N_t-1} matrices, each $N_\tau \times N_{f_d}$ cells in size and corresponding to a possible data bit combination. These matrices are incoherently added to the previous total incoherent integration to generate 2^{N_t-1} matrices of incoherent accumulations. But, only the matrix that corresponds to the most likely data combination at step l will be used as the new total incoherent integration. The other $2^{Nt-1} - 1$ matrices will be discarded. The most likely data combination is chosen as the one that maximizes the total incoherent integration at step l. This process is performed in parallel N_b times, producing N_b matrices of size $N_\tau \times N_{f_d}$.

3.2.4 Eliminating Doppler Bins

As the PIT increases, the number of the Doppler bins increases. This will cause an increase in the processing and memory overhead. To enable the use of long PIT with reduced processing, a Doppler bin elimination method is developed, which eliminates the unlikely Doppler bins after a predefined number of steps. This method can be justified based on the fact that the correct cell will always contain information due to the correlation between the received signal and the local one. If the correct cell does not produce the maximum power value, its value will be within a certain range of high values obtained from all of the cells. Thus, it is not necessary to keep track of all the Doppler bins throughout all of the processing steps.

The following variables are defined: Z is the number of times the Doppler bin elimination is performed; N_{ez} is the number of acquisition steps between the elimination steps $z-1$ and z; N_{elim} is the number of times the value of N_{ez} changes; N_{change} is a vector of size equal to N_{elim} that contains the acquisition step numbers after which the current N_{ez} is to be either increased or decreased. The amount of change in N_{ez} is defined in a vector $N_{ez_{change}}$ of N_{elim} size; E_{fz} is a fraction of the current Doppler bins to be eliminated; E_{change} is the amount of change in the current E_{fz}. The change in E_{fz} is to be made after each elimination step z; and E_{limit} is the maximum or minimum allowed value of E_{fz}.

After N_{ez} steps, a fraction of E_{fz} of the Doppler bins is eliminated. Then, E_{fz+1} is set equal to either: (1) $\max\{E_{limit}, E_{fz}+E_{change}\}$ if E_{change} is negative, or (2) $\min\{E_{limit}, E_{fz}+E_{change}\}$ if E_{change} is positive. The number of remaining bins after an elimination operation, N_{bin_z}, is not allowed to go below a predefined number of minimum, $N_{d_{min}}$ (i.e. $N_{bin_z} \geq N_{d_{min}}$, for $z = 1, \ldots, Z$). After each of the acquisition step numbers defined in N_{change}, N_{ez} is changed by an amount defined in $N_{ez_{change}}$.

The N_{bin_z} Doppler bins are selected out of the result from all of the possible bit edges, N_b, at once. This is done by identifying the cell that has the maximum power at each Doppler bin (i.e., identifying the code delay $\tau_{f_{d_v}}$ that produces the maximum power at each Doppler bin f_{d_v} for $v = 1, \ldots, N_{f_d}$). This is done for all N_b incoherent integration results. The $N_b N_{f_d}$ cells are then ordered according to their power values from the maximum to the minimum. Then, the highest N_{bin_z} powers are used to identify the Doppler bins that are preserved. If there are some repeated Doppler bin values among the chosen N_{bin_z} Doppler bins (each one will come from a different possible bit edge), then the algorithm keeps taking the following cell from the ordered group until there are N_{bin_z} distinct values of f_{d_v}. The other Doppler bins are eliminated and not included in subsequent integrations, and the incoherent accumulations resulting from the eliminated Doppler bins are also eliminated.

If more than one PIT is used in the acquisition, then E_{fz}, E_{change}, E_{limit}, and $N_{d_{min}}$ are vectors of size equal to the number of different PITs. Each element specifies a value for each PIT. Numerical values for the variables defined in this subsection have been determined by simulation at different C/N_0 values, with the goal of setting these values so as to make the probability of eliminating the correct cell approaches zero.

3.2.5 Increasing the Predetection Integration Time

Through periodic elimination of unlikely Doppler bins, as described above, it is possible to start with a small PIT and increase it without increasing the processing requirements. The PIT can be increased when the following condition is satisfied: $(\mu/(T_{I_g} - T_{I_{g-1}}))\, N_{bin_z} \leq N_{bin_{max}}$, in which $N_{bin_{max}}$ is the maximum allowed number of Doppler bins, T_{I_g} and $T_{I_{g-1}}$ are the gth and $(g-1)$th PIT lengths, and μ is a fraction used to determine the minimum separation between any two consecutive Doppler bins, where $0 < \mu \leq 1$. For example, at the gth PIT the minimum separation is μ/T_{I_g}.

The remaining Doppler bins, after increasing the PIT to T_{I_g}, will not, however, be a range of frequencies with constant separation due to the eliminated bins. New Doppler bins must be defined and spaced closer together to account for the narrower bandwidth resulting from the increase in the PIT. The frequencies of the new Doppler bins are calculated from the existing ones as follows:

(1) N_{DopRng} ranges are formed from the remaining Doppler bins of $T_{I_{g-1}}$. Each range contains Doppler bins with frequency separations of $\mu/T_{I_{g-1}}$ kHz. The separation between any two adjacent ranges is larger than $\mu/T_{I_{k-1}}$ kHz, however, as a result of the prior eliminations.

(2) Define $D_{g-1,i,nr}$ and $D_{g,j,nr}$ as the frequencies of, respectively, the ith Doppler bin in the nrth range of $T_{I_{g-1}}$ and the jth Doppler bin in the nrth range of T_{I_g}. $nr = 1, \ldots, N_{DopRng}$. The true Doppler shift could lie at a separation in frequency as large as $(0.5\,\mu/T_{I_{g-1}})$ from one of the ranges. For each of the N_{DopRng} ranges, the frequencies of the new Doppler bins of T_{I_g} are found by calculating the first and last Doppler bin frequencies, respectively, as

$$D_{g,1,nr} = D_{g-1,1,nr} - \frac{1}{2}\frac{\mu}{T_{I_{g-1}}}$$

$$D_{g,last,nr} = D_{g-1,N_{g-1,nr},nr} + \frac{1}{2}\frac{\mu}{T_{I_{g-1}}}$$

where $N_{g-1,nr}$ is the total number of the Doppler bins in the nrth range of $T_{I_{g-1}}$. The total number of the Doppler bins $N_{g,nr}$ in the nrth range of T_{I_g} can be calculated from

$$N_{g,nr} = \left\lceil \frac{T_{I_g}}{\mu} \left| D_{g,last,nr} - D_{g,1,nr} \right| \right\rceil + 1$$

(3) The frequencies of the rest of the Doppler bins of T_{I_g} are then found from

$$D_{g,i,nr} = D_{g,i-1,nr} + \frac{\mu}{T_{I_g}}, \qquad \text{for } i = 2, \ldots, \mathit{last}$$

Since some of the new Doppler bins at the gth PIT will not have an accumulation result up to the current step, their accumulations are set to the accumulation of the nearest old Doppler bins. If a new Doppler bin lies exactly in the middle of two old Doppler bins, its accumulation is set as the average of the accumulations belonging to the two old Doppler bins.

3.2.6 Implementing Circular Correlation with Multiple Data Bits

The CCMDB algorithm is implemented as two modules; one is the signal generator, and the other is the acquisition. The algorithm aims at finding the code delay at the start of the acquisition interval. Once the acquisition is concluded, the change in the code delay $\Delta\tau$ relative to the start of the acquisition interval can be approximated from the estimated Doppler shift $\hat{f}_d$ as

$$\Delta\tau = -T_t \frac{\hat{f}_d}{f_{L1}} \tag{3.6}$$

where T_t is the total acquisition time. For simplicity, the PIT T_{I_g} is set to be a multiple N_{t_g} of the exact data length (i.e., $T_{I_g} = 20\ N_{t_g}$), as opposed to a multiple of the Doppler-compensated data length. However, for each Doppler shift, Doppler-compensated I_L and Q_L local signals are generated at the time instances t_k, as in (3.3) and (3.4). The relative delay between the received and the local codes stays approximately the same during the acquisition. The only negative effect, therefore, of keeping the PIT an exact multiple of 20 ms is that the coherent integration will cross the bit edge transition. However, since the maximum acquisition time is only a few seconds, in the worst case, the start of the PIT will be off by only few samples from the bit edge transition; the number of those samples can be found from (3.6). In addition, since only N_b possible edge positions are considered, such a negative effect will be negligible compared to the loss due to the separation between the actual bit edge position and the nearest of the N_b possible edges. Figure 3.3 illustrates the implementation of the CCMDB algorithm.

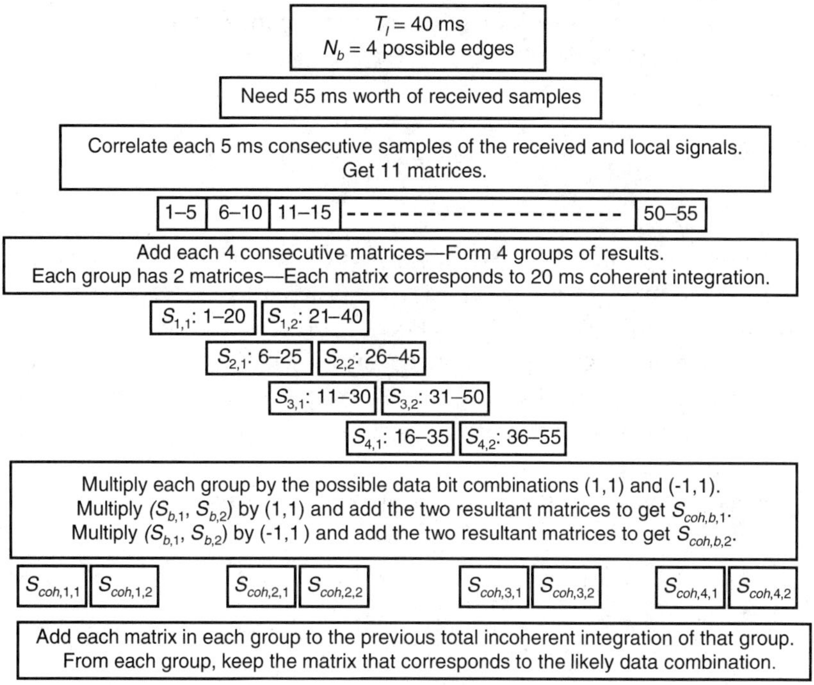

Figure 3.3 Illustration of the implementation of the CCMDB acquisition algorithm.

At each step, N_b coherent integrations are calculated, each starting at a possible bit edge position. These calculations are not independent, so they are done in such a way as to reduce the total processing requirements. First, the coherent integration, performed by circular correlation, is calculated for each T_{dms} ms. The calculation is done separately in blocks of T_{fft} ms, where $T_{fft} = T_{dms}/N_b$ is the separation between the possible bit edges.

The complex correlation between the received signal and the local one over each T_{dms} ms, within T_{I_g} ms, at a given bit edge b and at a step l of the acquisition, can be expressed as

$$s_{b,nt}(\tau_u, f_{d_v})\big|_l = \sum_{h=1}^{N_b} \sum_{n=1}^{N_h} r\left(t_{k_{h,n_t}+n-1}\right) \left[I_L\left(t_{k_{h,n_t}+n-1}, \tau_u, f_{d_v}\right) + j\, Q_L\left(t_{k_{h,n_t}+n-1}, \tau_u, f_{d_v}\right)\right]\Big|_l \tag{3.7}$$

where $b = 1, \ldots, N_b$ is a possible bit edge index; $nt = 1, \ldots, N_{t_g}$ is the index of each T_{dms} within one T_{I_g}; h is the index of each T_{fft} within one T_{dms}; n is the sample index within each T_{fft} ms; N_h is the number of samples in the T_{fft}; and

k_{h,n_t} is the index of the first sample in the hth T_{fft} ms of the current T_{dms} ms. Define the inner summation in (3.7) by $G_{h,\,b,\,n_t}$. So,

$$s_{b,\,n_t}(\tau_u,\,f_{d_v})|_l = \sum_{h=1}^{N_b} G_{h,\,b,\,n_t}(\tau_u,\,f_{d_v})|_l \tag{3.8}$$

It should be noted that the G term represents the coherent integration over T_{fft} ms. Each T_{fft} ms can be a part of up to N_b different s_{b,n_t} integrations. For example, $G(1,\,1,\,n_t) = G(2,\,N_b,\,n_t - 1)$, and $G(1,\,1,\,1)|_l = G(2,\,N_b,\,N_t)|_{(l-1)}$. Thus, the coherent integration over each T_{fft} ms is calculated only once, and then it is used in all the G values that represent the coherent integration over that same T_{fft} ms. The coherent integration for each T_{dms} ms, at each possible bit edge, can be found as follows: For the first T_{dms} in the first step (i.e., at $l = 1$), $s_{1,1}|_1$ can be found directly from (3.8). Then, $s_{b,\,n_t}|_l$ can be found from

$$\begin{aligned} s_{b,\,n_t}(\tau_u,\,f_{d_v})\big|_l = {} & s_{b-1,\,n_t}(\tau_u,\,f_{d_v})\big|_l - G_{1,b-1,\,n_t}(\tau_u,\,f_{d_v})\big|_l \\ & + G_{N_b,\,b,\,n_t}(\tau_u,\,f_{d_v})\big|_l \end{aligned} \tag{3.9}$$

for $b = 2, \ldots, N_b$; $nt = 1, \ldots, N_t$; and $l = 1, \ldots, L$; while $s_{1,\,nt}|_l$, for $nt = 2, \ldots, N_t$ and $l = 1, \ldots, L$, can be found from

$$\begin{aligned} s_{1,\,n_t}(\tau_u,\,f_{d_v})\big|_l = {} & s_{N_b,\,n_t-1}(\tau_u,\,f_{d_v})\big|_l - G_{1,N_b,\,n_t-1}(\tau_u,\,f_{d_v})\big|_l \\ & + G_{N_b,\,1,\,n_t}(\tau_u,\,f_{d_v})\big|_l \end{aligned} \tag{3.10}$$

$s_{1,1}|_l$, for $l = 2, \ldots, L$, can be found from

$$\begin{aligned} s_{1,\,1}(\tau_u,\,f_{d_v})\big|_l = {} & s_{N_b,\,N_{t_g}}(\tau_u,\,f_{d_v})\big|_{l-1} - G_{1,\,N_b,\,N_{t_g}}(\tau_u,\,f_{d_v})\big|_{l-1} \\ & + G_{N_b,\,1,\,1}(\tau_u,\,f_{d_v})\big|_l \end{aligned} \tag{3.11}$$

Following the above calculation, there will be N_{t_g} matrices of size $N_\tau \times N_{f_d}$ for each of the N_b possible bit edge positions. Each consecutive N_{t_g} of $s_{b,\,n_t}$ is then multiplied by the $2^{N_{t_g}-1}$ possible data combinations, and the results are added together to form $s_{coh_{b,E}}$. The index $E = 1, \ldots, 2^{N_{t_g}-1}$ refers to a possible data bit combination. So,

$$s_{coh_{b,E}}(\tau_u,\,f_{d_v})\Big|_l = \sum_{n_t=1}^{N_{t_g}} d_{E_{n_t}}\, s_{b,n_t}(\tau_u,\,f_{d_v})\Big|_l \tag{3.12}$$

where $d_{E_{n_t}}$ is a data value within N_{t_g}-length data and a possible data bit combination E. Each $s_{coh_{b,E}}$ is added incoherently to the previous accumulation to produce N_b groups, each containing $2^{N_{t_g}-1}$ matrices. From each group, only one matrix is used as the new total incoherent integration at the lth step, while the other matrices are discarded. The chosen matrix corresponds to the most likely data bit combination in the last T_{I_g}.

3.2.7 Determining the Most Likely Data Bit Combination

To determine the most likely data bit combination, the algorithm follows different approaches throughout the processing steps. This is to increase sensitivity and reduce degradation in the probability of detection due to an incorrect data bit combination. The general approach is as follows: Let $I_{coh_{b,E}}$ and $Q_{coh_{b,E}}$ define the real and imaginary parts, respectively, of (3.12). If $P_{b,l,E}$ defines the total incoherent integration at a step l for a possible bit edge b and data combination E, then

$$
\begin{aligned}
P_{b,l,E}(\tau_u, f_{d_v}) = {} & P_{b,l-1}(\tau_u, f_{d_v}) \\
& + \left[I_{coh_{b,E}}(\tau_u, f_{d_v})^2 + Q_{coh_{b,E}}(\tau_u, f_{d_v})^2 \right]_l
\end{aligned}
\tag{3.13}
$$

Each $P_{b,l,E}$ is a matrix of size $N_\tau \times N_{f_d}$. Only one of the 2^{N_t-1} of $P_{b,l,E}$ will form the new $P_{b,l}$, which will be the total integration at step l. This is chosen as the matrix that contains the maximum power, out of all cells in all matrices, in one of its cells. Define the cell that contains this maximum as $Cell(\tau_M, f_{d_N})_{max}$, where τ_M and f_{d_N} are, respectively, the code delay and Doppler shift that produced the maximum power. Then,

$$
P_{b,l} = P_{b,l,E}\big|_{Cell(\tau_M, f_{d_N})_{max} \in P_{b,l,E}}
\tag{3.14}
$$

The problem with applying this approach directly with weak signals is that the cell that produces the maximum power could be a noise cell. This means that the data combination is chosen randomly. At the beginning, any combination of the following alternatives is followed.

(1) A depth of D_i steps is defined in which the algorithm keeps track of all the matrices that correspond to all the possible data bit combinations. At each of the D_i steps, the new coherent matrices are added incoherently to the previous ones. This means that at the end of the D_i steps, there will be $2^{D_i(N_{t_g}-1)}$ matrices for each of the N_b edges, assuming that T_{I_g} does not change during the D_i steps. The matrix that has the maximum power

in one of its cells will be used to form the new incoherent integration for each N_b, and the other matrices will be discarded. In this case, $P_{b,l}$ will be calculated every D_i steps, where

$$P_{b,l} = P_{b,l-D_i} + \left\{ \sum_{n=1}^{D_i} I^2_{coh_{b,E,n}} + Q^2_{coh_{b,E,n}} \right\}_m \tag{3.15}$$

where m refers to the matrix, out of $2^{D_i(N_{tg}-1)}$ matrices, that contains the maximum power in one of its cells. Such an approach increases the storage space needed.

(2) Start with $N_{t_g} = 1$, and increase it with time. This will result in power accumulation in the correct cell and can reduce the probability of choosing an incorrect data bit combination.

(3) The $2^{N_{tg}-1}$ matrices are compared cellwise (i.e., cells with the same index in each matrix are compared), as opposed to choosing the $P_{b,lE}$ with the maximum power to be $P_{b,l}$. The maximum from each index forms the new incoherent integration at that index. At a delay τ_u and Doppler f_{d_v} the new result is

$$\begin{aligned} P_{b,l}(\tau_u, f_{d_v}) = \max \{ & P_{b,l1}(\tau_u, f_{d_v}), P_{b,l2}(\tau_u, f_{d_v}), \ldots, \\ & P_{b,l2^{N_{tg}-1}}(\tau_u, f_{d_v}) \} \end{aligned} \tag{3.16}$$

This alternative will cause some gain loss; it will also increase the power in all cells due to the maximization operation. However, there will be no loss due to an incorrect data bit combination in the correct cell. In addition, this alternative will give higher gain compared to using $N_{t_g} = 1$ for the same total integration time. To illustrate such an approach, a simulation test is done for a C/N_0 of 20 dB-Hz using T_{I_g} of 20 and 40 ms. Figure 3.4(a) shows the acquisition result for $T_{I_g} = 20$ ms after 60 steps (i.e., 60 data bits). Figure 3.4(b) shows the result for $T_{I_g} = 40$ ms after 30 steps (i.e., 60 data bits, the same amount of data as in the first case). As shown, using $T_{I_g} = 40$ ms results in a higher signal-to-noise ratio compared to using $T_{I_g} = 20$ ms.

(4) A different likely data bit combination is chosen at each code delay for all of the Doppler shifts. This data combination is chosen as the one that generates the maximum power among all the Doppler shifts in the $2^{N_{tg}-1}$ matrices at that code delay.

One or more of these four approaches can be followed for a number of steps equal to N_{max} before switching to the general one described above. If alternatives 3 and 4 are used, then the algorithm starts with alternative 3 and applies it for $N_{max\,All}$

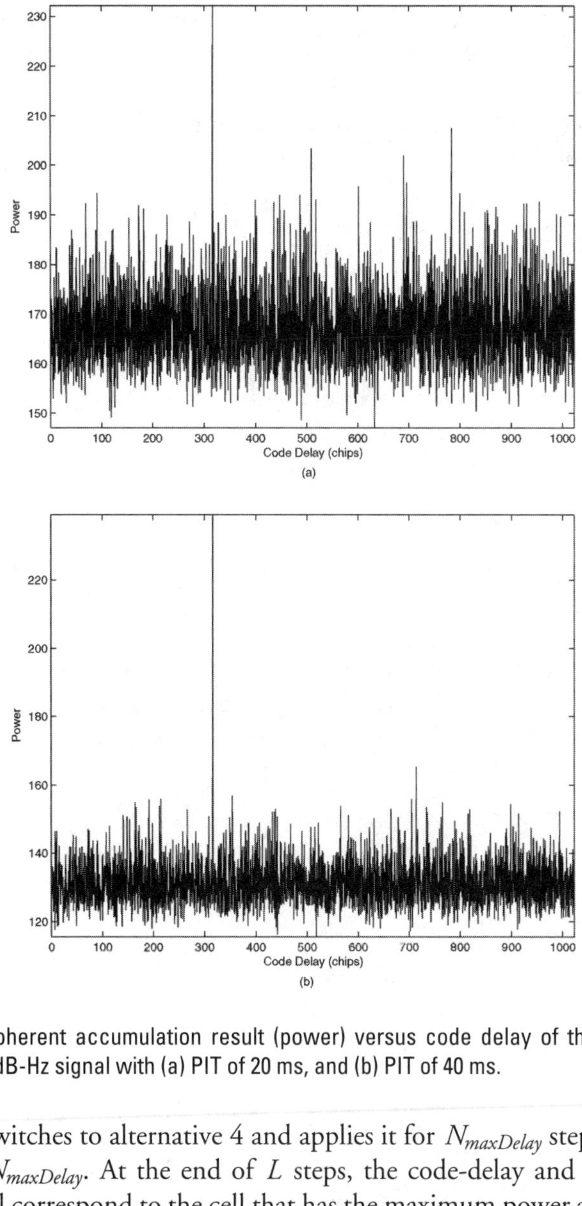

Figure 3.4 Incoherent accumulation result (power) versus code delay of the acquisition of 20-dB-Hz signal with (a) PIT of 20 ms, and (b) PIT of 40 ms.

steps, then switches to alternative 4 and applies it for $N_{maxDelay}$ steps; so, $N_{max} = N_{max All} + N_{maxDelay}$. At the end of L steps, the code-delay and Doppler shift estimates will correspond to the cell that has the maximum power among all cells in the N_b matrices, given that this maximum exceeds an acquisition threshold, γ.

3.2.8 Computational Analysis of the CCMDB

A separate circular correlation (FFT/IFFT) operation is done for each T_{fft} ms sample. For simplicity, assume the number of samples in each T_{fft} ms is constant and define it as N_H. The number of calculations for each circular correlation is proportional to $3N_H \log_2(N_H)$. For each possible Doppler shift, at the first step and at each step following an increase in the PIT, the number of circular correlation operations is $N_b N_{t_g} + N_b - 1$, while at the rest of the steps, this number is $N_b N_{t_g}$. Thus, the total amount of computation needed for the circular correlation throughout the acquisition process is

$$C_{cc} = 3N_H \log_2(N_H) \sum_{g=1}^{N_{T_I}} \left\{ N_b \sum_{j=1}^{L_g} N_{f_{gj}} N_{t_g} + N_{f_{g1}} (N_b - 1) \right\} \tag{3.17}$$

where $N_{f_{gj}}$ is the number of possible Doppler bins at the jth step of the gth PIT. L_g is the number of incoherent accumulations for the gth PIT. The coherent integration for each T_{dms}, except the first one, is calculated by one addition and one subtraction; its total amount of computation is

$$C_{ci} = 2N_b N_\tau N_{t_1} N_{f_{11}} + 2N_b N_\tau \sum_{k=1}^{N_{T_I}} N_{t_g} \sum_{j=1}^{L_g} N_{f_{gj}} \tag{3.18}$$

Both of the previous steps will generate $N_b N_{t_g}$ matrices with size equal to $N_\tau N_{f_{gj}}$ at the jth step of the gth PIT. Following that, the matrices that correspond to each possible data bit combination are generated. The total amount of addition/subtraction is

$$C_{dc} = N_b N_\tau \sum_{k=1}^{N_{T_I}} (N_{t_g} - 1)\, 2^{N_{t_g}-1} \sum_{j=1}^{L_g} N_{f_{gj}} \tag{3.19}$$

This will generate $N_b 2^{N_{t_g}-1}$ matrices with size $N_\tau N_{f_{gj}}$ after a step. The real and imaginary parts (representing the I and Q signal parts) are squared and added together. This amounts to a total computation of

$$C_{IQ} = 3N_b N_\tau \sum_{g=1}^{N_{T_I}} 2^{N_{t_g}-1} \sum_{j=1}^{L_g} N_{f_{gj}} \tag{3.20}$$

If the general method of choosing the most likely data bit combination is applied, then each matrix is added incoherently to the previous total incoherent integration; the computation needed at a step is $N_b N_\tau N_{f_{gj}} 2^{N_{t_g}-1}$. The matrix that corresponds to the most likely data combination at each possible bit edge position is kept, and the other $2^{N_{t_g}-1} - 1$ matrices are deleted. If the general method is used, finding the maximum can be done using a method such as that in [1].

Thus, the number of comparisons needed at a step is $N_b\,(2^{N_{tg}-1}\,N_\tau\,N_{f_{gj}}-1)$. If alternative 3 of Section 3.2.7 is used, then the number of comparisons at a step is $N_b\,2^{N_{tg}-1}\,N_\tau\,N_{f_{gj}}$. As can be seen, the computation and storage requirements are multiplied by the number of possible Doppler shifts. This number doubles with the PIT. Thus, Doppler bin elimination can largely reduce the processing and memory requirements and enable the use of a long PIT.

3.3 Algorithm 2: Modified Double Block Zero Padding

The DBZP calculates the coherent integration at all of the Doppler bins and all of the code delays in the same operation. Thus, it requires less processing as compared to the circular correlation. The problem is that only one replica code is used in the correlation calculation at all the Doppler bins. The replica code is not compensated by the Doppler effect on the code length. Thus, there will be a difference in the length between the received and the replica codes. This causes a difference in the start of the received code relative to the replica code at each consecutive T_I ms (i.e., a change in the code delay τ). Consequently, subsequent incoherent integrations will be added at different code delays relative to each other. This problem increases as the integration length increases. Thus, there is a limit to the maximum integration length, and, consequently, there is a limit to the minimum C/N_0 that can be acquired. The MDBZP circumvents the limitations of the DBZP, as explained in the next few subsections.

3.3.1 Compensating for Doppler Effects

The problems caused by the Doppler effect on the code length are handled in the MDBZP by dividing the whole Doppler range into a small number of ranges, N_{range}. Define f_{Rsize_i} as the size of the ith range, f_{mid_i} as the middle frequency of the ith range, and N_{fd_i} as the number of Doppler bins in the ith range. N_{range} replica code versions are generated. Each version's length is compensated for by the Doppler effect of one of the f_{mid_i}. The ith code version is used for the calculation of the coherent integration of the ith Doppler range. The objective of this is to reduce the relative change in the delay between the samples of the received and the replica codes over the T_I coherent integration (i.e., to reduce the difference in the amount of the code between each two corresponding samples of the received and the replica codes). If the true Doppler shift lies at the end of the ith range, then the relative change in the delay between the first and the last samples in the T_I ms is

$$\Delta\,\tau_i = T_I\,\frac{f_{Rsize_i}}{2\,f_{L1}} \tag{3.21}$$

Define the maximum allowable relative change in the delay, $\Delta\tau_{max}$, as a function in the sampling rate such that

$$\Delta\tau_{max} = \frac{1}{\lambda_{max} f_s} \tag{3.22}$$

where $\lambda_{max} \geq 1$ is a constant. The corresponding maximum range size is

$$f_{Rsize_{max}} = \frac{1}{\lambda_{max} f_s} \frac{2 f_{L1}}{T_I} \tag{3.23}$$

Let the total Doppler shift coverage be $\pm f_{dcov}$. The number of ranges is

$$N_{range} = \left\lceil \frac{2 f_{dcov}}{f_{Rsize_{max}}} \right\rceil \tag{3.24}$$

The DBZP is calculated N_{range} times, where each calculation uses one of the replica code versions. Each calculation produces a matrix of size $N_{f_d} \times N_\tau$ that represents the coherent integration for the whole Doppler range, not only the smaller ranges. However, only the coherent integration for the Doppler bins from O_{fs_i} to O_{fe_i} are taken from the matrix that is generated using the ith replica code, where

$$O_{fs_i} = \sum_{j=1}^{i-1} N_{fd_j} + 1 \tag{3.25}$$

$$O_{fe_i} = \sum_{j=1}^{i} N_{fd_j} \tag{3.26}$$

So, eventually, only one matrix of size $N_\tau \times N_{f_d}$ will be formed by combining these results, out of the N_{range} matrices.

Before this coherent integration is added incoherently to the previous total integration, another compensation is done. Since the replica code versions account for the change in the code length based on N_{range} of f_{mid_i} frequencies only, then if the true Doppler shift does not equal f_{mid_i}, the relative delay between the received and the replica codes will change after processing each T_I ms of data. The amount of this change depends on the difference between the true Doppler shift and the nearest f_{mid_i}. The compensation is done by circularly shifting the coherent integration result at each Doppler bin after processing each T_I ms of data, depending on the frequency difference. Figure 3.5 illustrates these steps.

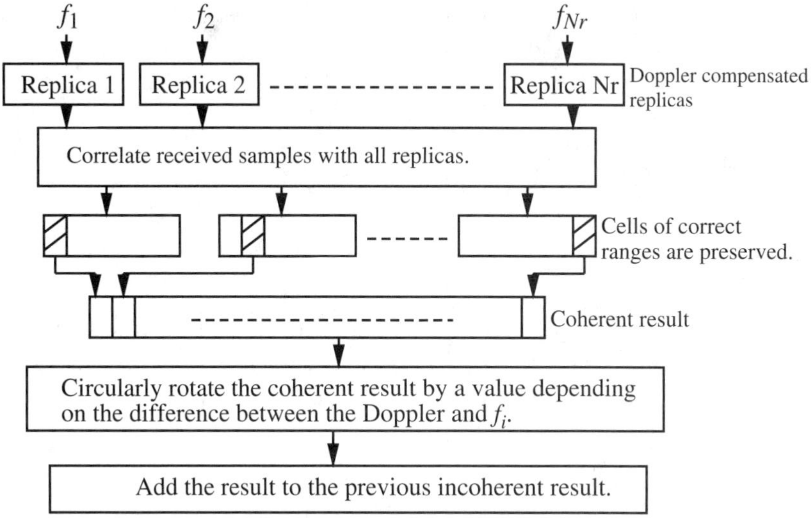

Figure 3.5 Illustration of the MDBZP acquisition algorithm's operation steps.

3.3.2 Handling Unknown Data Bits and Bit Edges

The algorithm uses a long coherent integration time, T_I, which is a multiple N_t of one data bit length, T_{dms}. Since the bit edge position is not known, N_b parallel incoherent integrations are maintained throughout the acquisition process. Each incoherent integration starts at a different possible bit edge position. The chosen N_b possible edge positions are less than 20, which is the total number of the possible positions. But the N_b possible positions are uniformly spaced over the 20 possible positions. As discussed in Section 3.2.2, if $N_b = 4$, then there will be a 10% loss in the integration. This loss will decrease as N_b increases. Also, the $2^{N_t - 1}$ possible data bit combinations will be tested as described in Section 3.2.7.

3.3.3 Implementing the MDBZP Algorithm

The MDBZP acquisition algorithm is implemented as follows. Assume first that the received signal in (2.1) is converted to the baseband to produce

$$r_c(t_k) = r(t_k)\, e^{-j 2\pi f_{IF} t_k} \tag{3.27}$$

The separation between each possible discrete code delay is set by the sample interval, $1/f_s$. Thus, the N_τ code delays are taken at every sample within one code period. The MDBZP produces Doppler bins with a frequency separation of

$f_{res} = 1/T_I$. Each T_I ms of data is divided into a number of blocks equal to the number of Doppler bins N_{f_d}:

$$N_{f_d} = 2\, f_{dcov}\, T_I \tag{3.28}$$

The size of each block is

$$S_{block} = \frac{N_\tau\, T_I}{N_{f_d}} \tag{3.29}$$

samples. The MDBZP calculates the correlation in steps. Each step generates a partial coherent integration for S_{block} possible code delays. The number of steps needed to calculate the correlation for all of the code delays is

$$N_{step} = \frac{N_\tau}{S_{block}} \tag{3.30}$$

The frequencies of the Doppler bins f_{d_v} can be calculated as

$$f_{d_v} = \left[v - (N_{f_d}/2) - 1\right] f_{res}, \quad v = 1, \ldots, N_{f_d} \tag{3.31}$$

The MDBZP accounts for N_b possible bit edge positions. In each step of the acquisition process, the MDBZP produces N_b sets of coherent integrations. Each set starts at a possible bit edge position. The N_b sets of coherent integrations are calculated from the same circular correlation blocks by shifting the bit edge by a multiple of S_{block} samples. Therefore, the possible bit edges will be aligned with some of the blocks resulting from the circular correlation. Two consecutive possible bit edge positions are separated from each other by T_{dms}/N_b ms or, equivalently, $N_\tau\, T_{dms}/N_b$ samples. This separation is equal to N_{bEdge} blocks, where

$$N_{bEdge} = \frac{N_\tau\, T_{dms}}{N_b\, S_{block}} = \frac{N_{f_d}}{N_t\, N_b} \tag{3.32}$$

In each step of the acquisition, the total data length needed to calculate the coherent integration for the N_b sets is $[T_I + (N_b - 1)(T_{dms}/N_b)]$ ms or, equivalently, $N_{tBlocks}$ blocks, where

$$N_{tBlocks} = N_{f_d} + (N_b - 1)\, N_{bEdge} \tag{3.33}$$

However, the first $(N_b - 1)\, N_{bEdge}$ blocks at the lth step of the acquisition are the same as the last $(N_b - 1)\, N_{bEdge}$ blocks at the $(l - 1)$th step. Because of this overlapping, at each step of the acquisition, except for the first step, the last N_{f_d} blocks of the $N_{tBlocks}$ are processed to produce a matrix of size equal to $N_{f_d} \times N_\tau$;

the result of processing the first $(N_b - 1)\ N_{bEdge}$ blocks is taken from the previous step. The processed blocks of the ith Doppler range are kept in a matrix M_{c_i} of size $(N_{f_d} + (N_b - 1)\ N_{bEdge}) \times N_\tau$, where in each acquisition step, at the ith Doppler range, after the processing of the N_{f_d} blocks, the last $(N_b - 1)\ N_{bEdge} \times N_\tau$ part of the matrix M_{c_i} is preserved in a matrix $M_{c\,Temp_i}$ to be used in the following step. In the first step of the acquisition, a total of $N_{tBlocks}$ blocks are processed. Figure 3.6 illustrates the implementation of the first step of the MDBZP, while Figure 3.7 illustrates the implementation of one step of the rest of the steps of the MDBZP. The algorithm works as follows:

(1) The received samples are divided into blocks. Every two adjacent blocks are combined into one block to produce N_{f_d} overlapping blocks, each with a size of $2S_{block}$.

(2) The local C/A code samples, with a length of T_I ms, are divided into N_{f_d} nonoverlapping blocks and each block, is padded with S_{block} zeros to produce N_{f_d} blocks, each of size equal to $2S_{block}$.

(3) At the m_rth step of the N_{step} steps, circular correlation is calculated between each two corresponding blocks of the received baseband signal samples and the locally generated C/A code samples. The circular correlation between each two blocks produces a block of size $2S_{block}$ points.

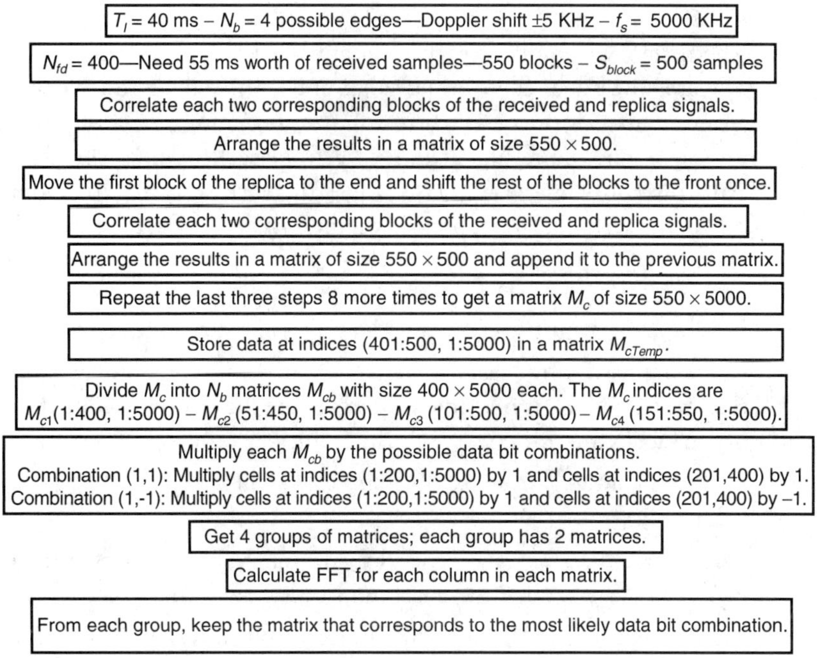

Figure 3.6 The implementation of the first step of the MDBZP acquisition algorithm.

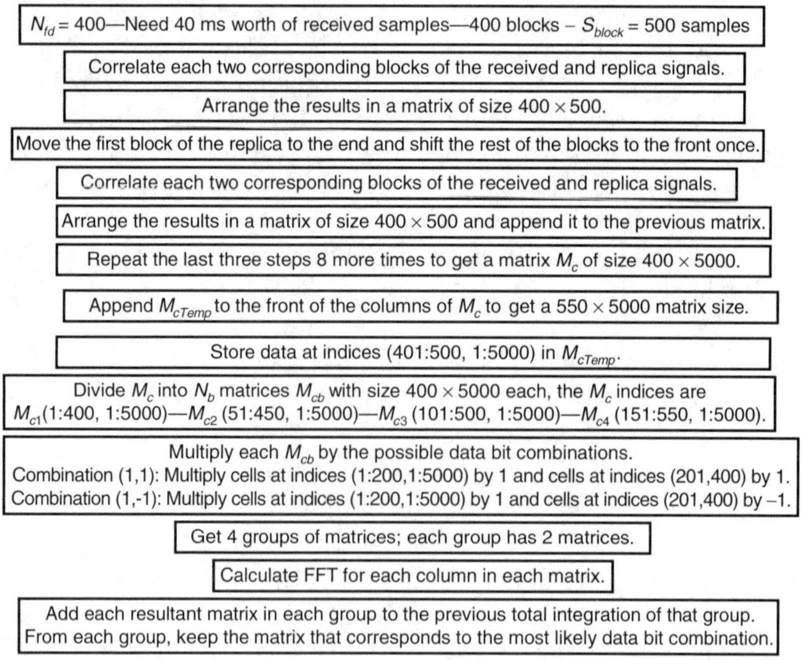

Figure 3.7 The implementation of a step of the MDBZP acquisition algorithm.

(4) The first S_{block} points of each resultant block are preserved, while the rest are discarded. The preserved points from all of the blocks are arranged in a form of a matrix of size $N_{fd} \times S_{block}$. The jjth column contains the results at index jj from each resultant block. The iith row contains the results of the iith block. For the ith Doppler range, these results are appended to the matrix M_{c_i} at the row indices from $[(N_b - 1)\ N_{bEdge} + 1]$ to $[(N_b - 1)\ N_{bEdge} + N_{fd}]$ and at column indices from $[(m_r - 1)\ S_{block} + 1]$ to $[m_r\ S_{block} + 1]$.

(5) Following each of the N_{step} steps, the blocks of the replica C/A code, which are used in the last step, are circularly shifted by one block such that the first block is moved back to become the last one, while the rest of the blocks are shifted once to the front.

(6) At the end of the N_{step} steps, the matrix M_{c_i} will have $N_{step}\ .\ S_{block} = N_\tau$ columns.

(7) The matrix $M_{c\,Temp_i}$ is appended to M_{c_i} at the row indices from 1 to $(N_b - 1)\ N_{bEdge}$ and at the column indices from 1 to N_τ.

(8) The elements of M_{c_i} at row indices from $[N_{fd} + 1]$ to $[N_{fd} + (N_b - 1)\ N_{bEdge}]$ and column indices from 1 to N_τ are preserved in the matrix $M_{c\,Temp_i}$ to be used in the following acquisition step.

(9) The matrix M_{c_i} is divided into N_b overlapping matrices. Each matrix $M_{c_{bi}}$ is of size equal to $N_{fd} \times N_\tau$, and represents the result of the bth possible bit edge position.

(10) Since each of the matrices $M_{c_{bi}}$, where $b = 1, \ldots, N_b$, is generated from a data of size equal to the coherent integration time T_I, where $T_I = N_t\ T_{dms}$, then each N_{f_d}/N_t rows of the $M_{c_{bi}}$ matrix can be viewed as generated using one of the N_t data bits. Thus, before performing the last step necessary to form the coherent integration for a total integration time of T_I, each $M_{c_{bi}}$ matrix is multiplied by one of the 2^{N_t-1} possible data bit combinations. This will generate 2^{N_t-1} matrices, for each of the N_b possible bit edges. Define these matrices as $M_{c_{Ebi}}$, where $E = 1, \ldots, 2^{N_t-1}$.

(11) The coherent integration is found by applying an FFT to each column of these $N_b\, 2^{N_t-1}$ matrices. Each cell (ii, jj) corresponds to the correlation at the Doppler bin ii and the code delay jj.

(12) The rows at the offsets O_{fs_i} to O_{fe_i} of the $M_{c_{Ebi}}$ matrix represent the coherent integration of the Doppler shifts of the ith Doppler range. Only these rows are appended to a matrix $M_{c_{Eb}}$ at offsets O_{fs_i} to O_{fe_i}. The other rows of the $M_{c_{Ebi}}$ matrix are discarded.

Steps (2) to (12) are repeated N_{range} times. Each repetition is done using one of the N_{range} replica code versions. At the end of the N_{range} processing, there will be $N_b\, 2^{N_t-1}$ matrices, each of which is of size $N_{f_d} \times N_\tau$ and represents a T_I coherent integration. Then, the algorithm continues as follows:

(1) A compensation is done to account for the change in code length resulting from the difference in the Doppler frequencies between the middle f_{mid_i} of each N_{range} and the other Doppler frequencies of each N_{range}. Since each row of $M_{c_{Eb}}$ corresponds to a possible Doppler frequency, it is circularly shifted by N_{s_v} samples, an amount that depends on the difference between f_{mid_i} and the frequency of the Doppler bin f_{d_v}:

$$N_{s_v} = \text{sign}(f_{d_v} - f_{mid_i})\ \text{round}\left[(l-1)\ T_I\ f_s\ \frac{|f_{d_v} - f_{mid_i}|}{f_{L1}}\right] \tag{3.34}$$

where l is the acquisition step number. The shifted rows form the coherent integration result at step l.

(2) For each possible bit edge position, only one matrix out of the 2^{N_t-1} matrices is preserved, while the other $2^{N_t-1} - 1$ are discarded. The preserved matrix corresponds to the most likely data bit combination, assuming a given bit edge position.

(3) Each of the N_b preserved matrices is added incoherently to the corresponding previous total accumulation. Thus, at the end of each step of the acquisition, there will be N_b accumulated matrices.

At the end of the L steps of the acquisition, if any cell in all of the N_b matrices has a value exceeding the threshold γ, then it will be concluded that a satellite has been detected. The cell that contains the maximum power will correspond to the estimated code delay and Doppler shift. The matrix that contains this maximum will correspond to the estimated bit edge position.

3.3.4 Computational Analysis

There are a total of L acquisition steps. At the first step, blocks of a total length of $[T_I + (N_b - 1)(T_{dms}/N_b)]$ ms are processed. For all subsequent steps, blocks of a total length of T_I are processed. In each of the acquisition steps, the processing of the blocks is done N_{steps} times. Each time, the blocks of the replica code are located at different shifts relative to the blocks of the received signal. The processing of each block will include calculating the FFT of each block of the received signal and the replica code, multiplying the FFT results of the corresponding blocks of the received signal and the replica code together, and finding the IFFT of the multiplication result. These three operations are repeated once for each of the N_{range} Doppler ranges. Throughout the N_{step} steps and the N_{range} Doppler ranges, the FFT is calculated only once for the received signal blocks; the FFT is calculated N_{range} times for the local code blocks; the multiplication is repeated N_{step} times for each of the N_{range} Doppler ranges; and the IFFT is calculated N_{step} times for each Doppler range. The total number of computations, C_{BP}, required by these operations is

$$C_{BP} = N_{f_d}\left[(1 + N_{range}(1 + N_{step}))\ 2S_{block}\ \log_2(2S_{block}) + 2\ S_{block}\ N_{range}\right] \tag{3.35}$$

The space requirement at the end of these operations is

$$S_{BP} = N_{range}\ N_\tau \left(N_{f_d} + 2(N_b - 1)\ \frac{N_{f_d}}{N_t\ N_b}\right) \tag{3.36}$$

using a unit of a matrix cell. Following that, each of the M_{c_i} matrices is divided into N_b matrices. Each of the $N_{range}\ N_b$ matrices is multiplied by the $2^{N_t - 1}$ possible data combinations. The computation due to this multiplication can be neglected since it is equivalent to only sign changes for the cells that need to be multiplied by a negative sign. The space needed, however, is

$$S_{DC} = 2^{N_t - 1}\ N_{range}\ N_b\ N_\tau\ N_{f_d} \tag{3.37}$$

Then, the FFT is calculated for each column of the generated matrices. Each FFT is performed over N_{f_d} points. Thus, the total number of computations is

$$C_{FFT} = 2^{N_t - 1}\, N_{range}\, N_b\, N_\tau\, N_{f_d}\, \log_2(N_{f_d}) \tag{3.38}$$

Then, the rows that correspond to the Doppler bins of each range are preserved, and the others are discarded. This will generate $N_b\, 2^{N_t-1}$ matrices of $N_\tau\, N_{f_d}$ size. The space needed will thus be

$$S_{AR} = 2^{N_t - 1}\, N_b\, N_\tau\, N_{f_d} \tag{3.39}$$

If the space needed is large, each range can be processed separately. This will reduce S_{BP} and S_{DC} by a factor of N_{range}.

The real and imaginary (I and Q) parts of each of the S_{AR} cells are squared and added together. This requires the following number of computations:

$$C_{IQ} = 3\, N_b\, 2^{N_t - 1}\, N_\tau\, N_{f_d} \tag{3.40}$$

If the general method of choosing the most likely data bit combination is applied, then each matrix is added incoherently to the previous total incoherent integration; the computation needed at a step is

$$C_{NC} = N_b\, 2^{N_t - 1}\, N_\tau\, N_{f_d} \tag{3.41}$$

The matrix that corresponds to the most likely data combination at each possible bit edge position is kept, and the other $2^{N_t-1} - 1$ matrices are deleted. If the general method is used, finding the maximum can be done using a method such as that in [1]. Thus, the number of comparisons needed is

$$C_{CM1} = N_b\, (2^{N_t - 1}\, N_\tau\, N_{f_d} - 1) \tag{3.42}$$

If alternative 3, described in Section 3.2.7, is used to choose the most likely data bit combination, then the number of comparisons at each step is

$$C_{CM2} = N_b\, 2^{N_t - 1}\, N_\tau\, N_{f_d} \tag{3.43}$$

Thus, after each step, only N_b matrices are kept. Each matrix has a size of $N_\tau \times N_{f_d}$. Each matrix holds the total incoherent integration up to that step. The major processing reduction in this algorithm is due to the fact that all the FFT and IFFT operations are performed over small number of points, $2\,S_{block}$ and N_{f_d}, as compared to the first algorithm, which performs FFT and IFFT separately for each possible f_d over the number of samples that span a multiple of one ms interval.

3.4 Signal Detection in the Presence of Strong Interfering Signals

Different C/A codes have nonzero cross-correlation. The cross-correlation between the local replica code and the code of a received interfering signal causes the interfering signal to appear with a reduced power from its original received power. The maximum reduced power is about 16.5 dB less than the original received power. This occurs if the true Doppler shift lies exactly on one of the Doppler bins, and there is no loss due to incorrect bit edge or data bit combination over a PIT that is larger than T_{dms}.

The interfering signal appears at its Doppler shift, f_d, and at each Doppler bin separated by a multiple of 1 kHz from f_d [2]. However, the cross-correlation exists at many code delays. Figure 3.8 shows the effect of the cross-correlation on the integration result if the local code does not exist in the received signal. Figure 3.8(a) shows the power versus the code delay. This figure plots only one value at each code delay, where this value is the maximum power from all the Doppler bins at that code delay. Figure 3.8(b) shows the power versus Doppler bins. It also plots only one value at each Doppler bin, where this value is the maximum power from all the code delays at that Doppler bin. This plotting approach is used throughout this chapter.

Two methods are developed to detect weak signals in the presence of strong interfering signals. The first method deals with the case when the received power

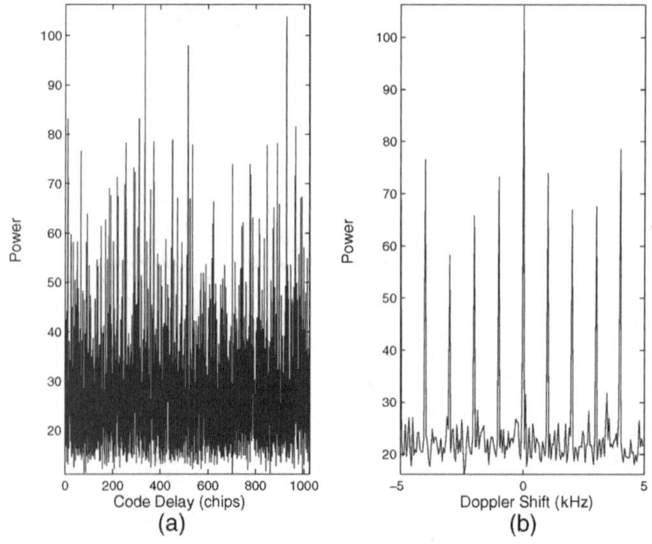

Figure 3.8 Cross-correlation effect on the integration result. The local code does not exist in the received signal: (a) power versus code delay, and (b) power versus Doppler shift.

of the interfering signal is at most 15 dB higher than the power of the expected signal (i.e., the signal that needs to be acquired); the second method deals with the case when there are higher interfering powers. Both of these methods are different from the approach followed in [3, 4], where the strong interfering signals are detected and tracked before the weak signals are detected. In such an approach, the tracking output is used to subtract the strong signals from the received signal before the correlation is calculated.

3.4.1 The Interfering Power Is Less Than the Expected Signal Power

In this case, the acquisition algorithm is the same as those described in Section 3.2 and Section 3.3, except that the total number of integrations L and the acquisition threshold γ are designed based on the noise and the interfering signal. The probability of false alarm p_f is calculated as

$$p_f = \max\{p_{f_n}, p_{DC}\}$$

where p_{f_n} is the probability of false alarm if there are no interfering signals, and p_{DC} is the probability of detecting of the interfering signal. γ is calculated from

$$\gamma = \max\{\gamma(p_{f_n}), \gamma(p_{DC})\}$$

where $\gamma(p_{f_n})$ is the threshold calculated based on the noise power and p_{f_n}. $\gamma(p_{DC})$ is the maximum expected power level of the interfering signal.

To be able to use a PIT multiple N_t of T_{dms}, the most likely data bit combination is estimated in each cell independently of the other cells, as described in Section 3.2.7, throughout the processing steps, so $N_{max} = L$. Also, an alternative method is to choose one most likely data combination for each Doppler bin at all of its code delays. This will be the combination that generates the maximum power for that Doppler bin.

Both p_f and γ are calculated taking into consideration the correct PDF resulting from such an approach. The PDF and p_f are calculated numerically because it is not possible to calculate them in a closed form; this is described in Section 3.6.

3.4.2 The Interfering Power Is Greater Than or Equal to the Expected Signal Power

This method takes advantage of the integration characteristics of the interfering signals shown in Figure 3.8. It can work only if the Doppler shift of the strong signal does not affect the Doppler shift of the correct signal (i.e., the Doppler shift of the interfering signal does not appear on the same bin as the Doppler shift of the expected signal). In this case, the threshold is designed based on the noise only. Also, the most likely data bit combination is estimated in each cell independently of other cells, as described in Section 3.2.7, throughout the processing steps, so $N_{max} = L$.

A test is performed to determine whether a signal is present. This test is based on the observation that the interfering signal will cause many code delays to exceed the threshold at each of the interfering Doppler bins, while only the correct signal will have only one code delay that exceeds the threshold at its correct Doppler bin. The test works as follows:

(1) Find all Doppler bins that exceed the threshold.
(2) If more than one code delay exceeds the threshold at the same Doppler bin, discard that Doppler bin.
(3) The detected signal will be concluded only if there is a Doppler bin that has only one code delay above the threshold.

A combination of the above two approaches is used if there are many interfering signals with different powers.

3.5 High Dynamic Acquisition

3.5.1 The Doppler Rate Effect on the Acquisition

If the receiver is accelerating, the Doppler shift will change with a rate depending on the relative line of sight between the satellite and the receiver. A change in the Doppler shift causes a change in the code length.

A Doppler bin f_{d_V} will contain a fraction β of the acquired signal power if the acquired Doppler shift f_d is located less than $1/\ T_I$ kHz from that bin, where

$$\beta = \text{sinc}^2\left([f_d - f_{d_V}]\ T_I\right)$$

A signal can be acquired without considering the Doppler rate if the Doppler shift does not move from one bin to another one located at a separation larger than $1/\ T_I$ KHz from the original one during the total integration time. This can be expressed by the following condition:

$$f_d + \alpha\, T_t < f_d + \frac{1}{T_I} \tag{3.44}$$

or, equivalently,

$$\alpha < \frac{1}{T_I\, T_t} \tag{3.45}$$

where T_I is the PIT, and T_t is the total integration time. However, if the Doppler shift of the received signal changes during the acquisition process, the signal power will scatter over more than one Doppler bin. This means that a signal can be acquired without considering the Doppler rate if the total power accumulated in the closest bin crosses the acquisition threshold γ during T_t. Weak-signal acquisition needs long T_I and T_t, which puts a limit on the maximum acquired Doppler rate.

Let f_{d_l} define the received Doppler shift at the start of the lth interval and $f_{d_{avl}}$ define the average received Doppler shift over the lth interval, where

$$f_{d_{avl}} = f_{d_l} + \alpha\, \frac{T_I}{2} \tag{3.46}$$

assuming a constant acceleration. In this case, a Doppler bin f_{d_V} will contain a fraction β_l of the acquired signal power in the lth step if $f_{d_{avl}}$ is separated from f_{d_V} by a value less than $1/T_I$ KHz. β_l can be approximated by

$$\beta_l \approx \text{sinc}^2([f_{d_{avl}} - f_{d_V}]\, T_I) \tag{3.47}$$

If $f_{d_{avl}}$ is located n_f / T_I KHz from f_{d_V}, $n_f \in [0, 1]$, then there are two cases to consider:

(1) If $f_{d_{avl}}$ is moving toward f_{d_V}, then the signal power will continue to accumulate in that bin until $f_{d_{avl}}$ passes through f_{d_V} and arrives at a distance of $1/T_I$ KHz from f_{d_V} or until $f_{d_{avl}}$ changes by an amount of $(n_f + 1)/T_I$ KHz. This will take $(n_f + 1)/(|\alpha|\, T_I)$ ms or, equivalently, a number of accumulation steps equal to

$$N_s = \left\lfloor \frac{n_f + 1}{|\alpha|\, T_I^2} \right\rfloor + 1$$

(2) If $f_{d_{avl}}$ is moving away from f_{d_V}, then the signal power will continue to be contained in that bin for about $(1 - n_f)/(|\alpha|\, T_I)$ ms or, equivalently, for a number of accumulation steps equal to

$$N_s = \left\lfloor \frac{1 - n_f}{|\alpha|\, T_I^2} \right\rfloor + 1$$

After each of the N_s steps, fd_{avl} will change by an amount of $\Delta f = \alpha\ T_I$. This means that n_f will change by an amount of $\Delta n_f = \alpha\ T_I^2$. At the end of the N_s steps, the total power in fd_V can be calculated from

$$P_{fd_V} = \sum_{l=1}^{N_s} p_l\ \beta_l \tag{3.48}$$

$$P_{fd_V} \approx \sum_{l=1}^{N_s} p_l\ \text{sinc}^2([fd_{avl} - fd_V]\ T_I) \tag{3.49}$$

$$P_{fd_V} \approx \sum_{l=1}^{N_s} p_l\ \text{sinc}^2([fd_{av1} + (l-1)\ \alpha\ T_I - fd_V]\ T_I) \tag{3.50}$$

where p_l is the signal power at the lth step, and

$$N_s = \left\lfloor \frac{(fd_V - fd_{av1})\ T_I}{\alpha\ T_I^2} + \frac{1}{|\alpha|\ T_I^2} \right\rfloor + 1 \tag{3.51}$$

It can be concluded that the maximum total power $P_{fd_V\ max}$ in fd_V will result when fd_{avl} starts at a distance of $1/\ T_I$ from fd_V, passes through fd_V, and reaches a distance of $1/\ T_I$ from fd_V in the other direction. This will require a number of steps $N_{s\,max}$ equal to

$$N_{s\,max} = \left\lfloor \frac{|fd_{av1} - fd_V|\ T_I + 1}{|\alpha|\ T_I^2} \right\rfloor + 1 \tag{3.52}$$

Thus, a signal can be acquired without considering α if a bin fd_V can accumulate a power above γ during the move of fd_{avl} from $fd_{V\pm1}$ to $fd_{V\mp1}$, the two bins adjacent to fd_V. Thus, the condition in (3.45) is equivalent to

$$P_{fd_V} \geq \gamma \tag{3.53}$$

If the Doppler bins frequency separation is μ/T_I, where $0 < \mu \leq 1$, then define fd_{Cl} as the bin closest to fd_{avl} at the lth step, which is the bin that has the maximum β_l at that step, and fd_{Ml} as the bin that contains the maximum power, which is calculated as in (3.48), at the lth step. If the maximum expected Doppler rate is α_{max}, then fd_{Cl} will change from being a reference to fd_V to being a reference to an adjacent bin after, at most, a number of steps equal to

$$S_c = \left\lfloor \frac{\mu}{|\alpha_{max}|\ T_I^2} \right\rfloor + 1 \quad \text{steps}$$

This is the case if $f_{d_{avl}}$ starts in the middle of f_{d_V} and an adjacent bin and ends in the middle of f_{d_V} and the other adjacent bin. However, in order for $f_{d_{Ml}}$ to change its reference from f_{d_V} to an adjacent bin, the total power in the adjacent bin has to exceed the total power in f_{d_V}.

3.5.2 High Dynamic Acquisition Algorithm

The objective of this algorithm is to identify the most likely $f_{d_{Cl}}$ after each step l of the total acquisition steps L and to accumulate the power of those $f_{d_{Cl}}$ together. If $P_{t_{CL}}$ defines the total power of the closest bins in all the L steps, and $P_{f_{dCl}}$ defines the power in the closest bin at the lth step, then

$$P_{t_{CL}} = \sum_{l=1}^{L} P_{f_{dCl}} \tag{3.54}$$

If the acquisition is concluded, the estimated Doppler shift will be defined by $f_{d_{CL}}$, which is the closest bin at the last step.

The most likely $f_{d_{Cl}}$ is found by a test. However, since $f_{d_{Cl}}$ will refer to the same Doppler bin for at most S_c steps, then it is not necessary to perform that test after each step l. Since the location of the initial $f_{d_{avl}}$ relative to $f_{d_{Cl}}$ is not known, a limit is established on the maximum number of allowed consecutive accumulations between each test as

$$S_m = \left\lceil \frac{S_c}{4} \right\rceil \tag{3.55}$$

The Doppler rate can be positive or negative; hence, the algorithm works as follows: An incoherent accumulation is performed for the first S_m steps using either of the two algorithms developed earlier. The result is stored as P_1, which is a matrix of size $N_\tau \times N_{f_d}$. Then, for the next S_m steps, a new incoherent accumulation $R_{S_m,1}$ is also calculated over these S_m steps. Before finding any new accumulation, the algorithm performs three additions using these two accumulations. The first one assumes that $f_{d_{Cl}}$ still refers to the same Doppler bin during the last S_m steps as the previous S_m steps. The second and third ones assume that $f_{d_{Cl}}$ has changed its reference to either of the adjacent bins due to a negative and a positive α, respectively. The three additions are

$$Q_{t,1}(\tau_u, f_{d_v}) = R_{S_m,1}(\tau_u, f_{d_v}) + P_1(\tau_u, f_{d_v}) \tag{3.56}$$

$$Q_{t,2}(\tau_u, f_{d_v}) = R_{S_m,1}(\tau_u, f_{d_v}) + P_1(\tau_u, f_{d_{v+1}}) \tag{3.57}$$

$$Q_{t,3}(\tau_u, f_{d_v}) = R_{S_m,1}(\tau_u, f_{d_v}) + P_1(\tau_u, f_{d_{v-1}}) \tag{3.58}$$

Then, the new total accumulation is found by comparing the three Q_t matrices at each cell and taking the maximum value. So,

$$P_2(\tau_u, f_{d_v}) = \max\{Q_{t,1}(\tau_u, f_{d_v}), Q_{t,2}(\tau_u, f_{d_v}), Q_{t,3}(\tau_u, f_{d_v})\} \quad (3.59)$$

After that, a new accumulation $R_{S_m,2}$ is performed over the next S_m steps, and then (3.56) to (3.59) are repeated. The algorithm follows the same steps until an acquisition decision is concluded.

There is also the effect of the Doppler shift on the code length. This effect will cause the signal power to appear at different code delays. However, because both acquisition algorithms consider this effect, the Doppler bins will be added at the correct code delay, so no additional steps are needed.

After the acquisition is concluded, an approximate Doppler rate is determined as follows. The power of the Doppler bins adjacent to the acquired Doppler shift is inspected to determine whether α is positive or negative. This is because the algorithm accumulates the power in the correct bins throughout its processing steps. After (3.59) is calculated at the s_lth step, where $s_l = l/S_m$, given that f_{dCl} changes its reference from a bin f_{dV} to the adjacent one f_{dV+1}, then the power in both bins at that step are related by

$$P_{s_l}(\tau_U, f_{dV+1}) \approx P_{s_l-1}(\tau_U, f_{dV}) + R_{S_m,s_l-1}(\tau_U, f_{dV+1}) \quad (3.60)$$

where τ_U is the code-delay estimate at the start of the search. It is not necessary to determine whether α is positive or negative in a separate step, although determining the sign of α will reduce the processing incurred in determining the value of α. Following that, the algorithm continues with the accumulation, but it calculates it only at the correct code delay and at few adjacent bins to the acquired Doppler shift. Then, by observing the time it takes the signal to appear at an adjacent Doppler bin, an approximate α can be calculated. If f_{d_1} and f_{d_2} define the Doppler bins with the maximum power at time t_1 and t_2, respectively, then an approximate α can be estimated from

$$\hat{\alpha} \approx \frac{f_{d_2} - f_{d_1}}{t_2 - t_1} \quad (3.61)$$

3.6 Probabilities of Detection and False Alarm

Following from (2.19) and (2.20), the result of a 1-ms circular correlation between the received signal of a stationary receiver and a local signal can be modeled as

$$I_i(\tau_u, f_{d_v}) = A\, d_i\, R(\tau - \tau_u)\ \mathrm{sinc}((f_d - f_{d_v})\, T_I)\ \cos(\theta_{e_i}) + n_{I_i} \quad (3.62)$$

$$Q_i(\tau_u, f_{d_v}) = A\, d_i\, R(\tau - \tau_u)\ \mathrm{sinc}((f_d - f_{d_v})\, T_I)\ \sin(\theta_{e_i}) + n_{Q_i} \quad (3.63)$$

where τ_u and fd_v are a possible code delay and a possible Doppler shift, respectively; τ is the correct code delay; f_d is the correct Doppler shift; d is a data bit value; $R(.)$ is the autocorrelation function of the C/A code; T_1 is the 1-ms correlation interval; θ_e is the average phase error over the integration interval; A is the signal level, which is normalized to derive the noise variance to 1; and n_I and n_Q are WGN with zero mean and unit variance.

3.6.1 Total Integration

The calculation of the probabilities of false alarm p_f and detection p_d are computed, taking into consideration all of the approaches used in the algorithms. An equivalent total integration is used, which accounts for the probability of choosing an incorrect data bit combination. An incorrect data bit combination causes a reduction in the signal power. Each data bit with an incorrect sign will cancel out another one with a correct sign, within 1 T_I ms. This results in signal averaging over $N_t - 2n_e$ data bit intervals, where n_e is the number of data bits with incorrect signs in T_I ms. This means that the number of possible values that could result from a coherent integration that is a multiple N_t of one data bit interval is

$$N_{av} = \left\lfloor \frac{N_t}{2} \right\rfloor + 1 \tag{3.64}$$

If N_t is an even number, then a "no signal" averaging is possible. Since the acquisition algorithm uses more than one PIT, define the following: N_{T_I} is the number of PITs used. N_{t_g} is the number of T_{dms} in the gth PIT. M and M_g are the number of code periods in, respectively, T_{dms} and the gth PIT. N_{av_g} is the number of possible results of signal averaging using the gth PIT. L_{gh} is the total number of incoherent integrations in the gth PIT and the hth result of N_{av_g}. L_g and L are the total number of incoherent integrations, respectively, in the gth PIT and in the whole acquisition process. So,

$$L_g = \sum_{h=1}^{N_{av_g}} L_{gh}$$

$$L = \sum_{g=1}^{N_{T_I}} L_g \tag{3.65}$$

There is a probability of choosing an incorrect data bit combination, where this probability depends on the acquired C/N_0. Define p_{gh} as the probability of choosing the data bit combination in the gth PIT that produces the hth result out of N_{av_g}. So,

$$\sum_{h=1}^{N_{av_g}} p_{gh} = 1$$

$$L_{gh} = p_{gh} L_g \tag{3.66}$$

The total integration is expressed by

$$P(\tau_u, f_{d_v}) = \sum_{g=1}^{N_{T_I}} \left[\frac{1}{M_g} \sum_{h=1}^{N_{av_g}} \left[\sum_{j=1}^{L_{gh}} \left[\left(\underbrace{\sum_{i=e_{ghj}}^{e_{ghj}+M_g-1} I_i(\tau_u, f_{d_v})}_{\{1\}} \right)^2 + \left(\underbrace{\sum_{g=e_{ghj}}^{e_{ghj}+M_g-1} Q_i(\tau_u, f_{d_v})}_{\{2\}} \right)^2 \right]_j \right]_h \right]_g \tag{3.67}$$

where e_{ghj} is the index of the first 1 ms of the coherent integration that corresponds to the jth incoherent integration of the hth possible signal averaging and the gth PIT. The division by $1/M_g$ is used to maintain the noise variances of the I and Q signals equal to 1.

3.6.2 Probability of False Alarm p_f

The probability of false alarm p_f specifies the probability that (3.67) will produce a power value higher than the acquisition threshold γ if there is no signal present. When the PIT T_I is a multiple N_t of T_{dms} ms, and the general approach for choosing the most likely data combination is applied, then the algorithm does the following: it multiplies the N_t matrices of the T_{dms}-ms coherent integration by the 2^{N_t-1} possible data bit combinations; it adds each N_t matrices of the same data bit combination; it adds the resultant 2^{N_t-1} matrices to the previous incoherent integration; then, it chooses the matrix that has the maximum power in one of its cell as the new incoherent integration. This means that at a step l, the new incoherent integration of the cell that has the maximum power is determined from

$$P_l(\tau_M, f_{d_N}) = P_{l-1}(\tau_M, f_{d_N}) + \max_j \left\{ \left(\sum_{m=1}^{N_t} d_{(j,m)} \; I_m(\tau_M, f_{d_N}) \right)^2 + \left(\sum_{m=1}^{N_t} d_{(j,m)} \; Q_m(\tau_M, f_{d_N}) \right)^2 \right\}_l \tag{3.68}$$

where P_l is the incoherent integration at step l; $d_{(j,m)}$ is the mth data bit value (out of N_t values) of the jth possible data bit combination (out of 2^{N_t-1} combinations); I_m and Q_m are the coherent integration result over the mth data bit interval within the T_I ms; and τ_M and f_{d_N} are, respectively, the code delay and the Doppler bin that generated the maximum power. Define J_l as the data bit combination that generated the maximum power in (3.68) in the lth step. The other cells at the lth step will have the new incoherent integration calculated using J_l; this is expressed by

$$P_l(\tau_u, f_{d_v}) = P_{l-1}(\tau_u, f_{d_v}) + \left[\left(\sum_{m=1}^{N_t} d_{(J_l,m)} \; I_m(\tau_u, f_{d_v}) \right)^2 + \left(\sum_{m=1}^{N_t} d_{(J_l,m)} \; Q_m(\tau_u, f_{d_v}) \right)^2 \right]_l \tag{3.69}$$

The exact derivation of the PDF of the integration is complicated because the cell and the data combination that produced the maximum at the lth step are not necessary those that produced the maximum at the $(l-1)$th step. Also, because the maximization in (3.68) is taken over two separate, independent functions (I and Q), this maximization is not equivalent to

$$P_l(\tau_M, f_{d_N}) = P_{l-1}(\tau_M, f_{d_N}) + \left[\left(\sum_{m=1}^{N_t} |I_m(\tau_M, f_{d_N})| \right)^2 + \left(\sum_{m=1}^{N_t} |Q_m(\tau_M, f_{d_N})| \right)^2 \right]_l \tag{3.70}$$

If the most likely data bit combination is chosen in each cell independently of the other cells for a total of N_{max} steps, as described in Section 3.2.7 and (3.16), then during these steps, the accumulation is determined from (3.68) for all the cells, not only for one cell.

In the following derivations, boldface capital letters are used to express a random variable, while small letters are used to express a value.

3.6.3 p_f Without Data Combination

Assume that the effect caused by multiplying by a possible data bit combination is not considered. In this case, the two summations over N_{av_g} and L_{gh} in (3.67) will be replaced by one summation over L_g, so that

$$P(\tau_u, f_{d_v}) = \sum_{g=1}^{N_{T_I}} \left[\frac{1}{M_g} \sum_{j=1}^{L_g} \underbrace{\left[\left(\underbrace{\sum_{i=e_{gj}}^{e_{gj}+M_g-1} I_i(\tau_u, f_{d_v})}_{\{1\}} \right)^2 + \left(\underbrace{\sum_{g=e_{gj}}^{e_{gj}+M_g-1} Q_i(\tau_u, f_{d_v})}_{\{2\}} \right)^2 \right]_j}_{\{3\}} \right]_g \tag{3.71}$$

where e_{gj} is the index of the first 1 ms of the coherent integration that corresponds to the jth incoherent integration of the gth PIT. In (3.71), since the noise is normally distributed, the outputs of {1} and {2} are also normally distributed. Define **X** as the amplitude of the noise in {3}; **X** has a Rayleigh distribution. The Rayleigh distribution has the following form

$$f_{\mathbf{X}}(x) = \frac{x}{\sigma_n^2} e^{-\frac{x^2}{2\sigma_n^2}} \tag{3.72}$$

Since part {3} of (3.71) is the square of the noise (with $\sigma_n^2 = 1$), then define it as **Y**; its distribution, $f_{\mathbf{Y}}(y)$, is found, as in [5], by

$$f_{\mathbf{Y}}(y) = \frac{1}{2\sqrt{y}} \left[f_{\mathbf{X}}\left(\sqrt{y}\right) + f_{\mathbf{X}}\left(-\sqrt{y}\right) \right] \tag{3.73}$$

This leads to the following PDF:

$$f_{\mathbf{Y}}(y) = \frac{1}{2} e^{-\frac{y}{2}} \tag{3.74}$$

Since L_g of **Y** are added together to form the incoherent integration, then the PDF of the result is the convolution of L_g of the PDF $f_{\mathbf{Y}}(y)$. Define the addition result of the L_g of **Y** as **Z**. The convolution is found by taking the Laplace transform of each of the $f_{\mathbf{Y}}(y)$, multiplying the results together, then taking the inverse Laplace transform. This leads to the following PDF:

$$f_{\mathbf{Z}}(z) = \frac{1}{2^{L_g}} \frac{z^{L_g - 1}}{(L_g - 1)!} e^{-\frac{z}{2}} \tag{3.75}$$

Since N_{T_I} of such L_g are added together, define the addition result as **S**. The PDF of the addition of all the L_g is

$$f_{\mathbf{S}}(s) = \frac{1}{2^L} \frac{s^{L-1}}{(L-1)!} e^{-\frac{s}{2}} \tag{3.76}$$

From (3.75) and (3.76), it is concluded that the PDF of the noise does not depend on the PIT length, but it depends on the total number of the incoherent accumulations. This holds as long as the noise variance is normalized to 1 in each PIT, which is achieved by dividing the integration by M_g, as in (3.71). The probability of false alarm is

$$p_f = \int_{\gamma}^{\infty} f_{\mathbf{S}}(s)\, ds \tag{3.77}$$

The integration of (3.77) gives

$$p_f = \left[1 + \frac{\gamma}{2} + \frac{1}{2!}\left(\frac{\gamma}{2}\right)^2 + \ldots + \frac{1}{(L-1)!}\left(\frac{\gamma}{2}\right)^{L-1}\right] e^{-\frac{\gamma}{2}} \tag{3.78}$$

Since at the last step the acquisition is concluded from the cell that has the maximum power among all the cells in all the N_b accumulations, then the distribution of the maximum $[\Lambda = \max(\mathbf{S}_1, \mathbf{S}_2, \ldots, \mathbf{S}_{N_{cell} N_b})]$, where N_{cell} is the total number of cells at the last step (this number depends on the last PIT used and the uneliminated Doppler bins), is

$$F_{\Lambda}(\lambda) = F_{\mathbf{S}_1, \mathbf{S}_2, \ldots, \mathbf{S}_{N_{cell} N_b}}(\lambda, \lambda, \ldots, \lambda) \tag{3.79}$$

A bound on the distribution of such Λ is established in [6]. The bound applies whether the $\mathbf{S}_i$'s are independent or not, where the index i refers to a cell out of the total number of cells. If $F_{\mathbf{S}_i}$ defines the marginal distribution of $\mathbf{S}_i$, then the bound is written as

$$\max\left(0,\ 1-\sum_{i=1}^{N_{cell}N_b}\bar{F}_{\mathbf{S}_i}(\lambda)\right)\leq F_\Lambda(\lambda)\leq\min_i F_{\mathbf{S}_i}(\lambda) \tag{3.80}$$

where $\bar{F}=1-F$. p_f in (3.77) is equivalent to

$$p_f=\bar{F}_\Lambda(\gamma) \tag{3.81}$$

Using the bound in (3.80), the following bound is established:

$$\bar{F}_\Lambda(\gamma)\leq\sum_{i=1}^{N_{cell}N_b}\bar{F}_{\mathbf{S}_i}(\gamma) \tag{3.82}$$

If all the $\mathbf{S}_i$'s have the same distribution, defined as $F_{\mathbf{S}}$, then

$$p_f=\bar{F}_\Lambda(\gamma)\leq N_{cell}\,N_b\,\bar{F}_{\mathbf{S}}(\gamma) \tag{3.83}$$

3.6.4 p_f with Data Combination

Assume that the most likely data bit combination is chosen for each cell, independently of the other cells, for a total of $N_{max\,All}$ steps. Then, a different data bit combination is chosen at each code delay for a total of $N_{maxDelay}$ steps; $N_{max}=N_{max\,All}+N_{maxDelay}$. Then, the general approach is followed for the rest of the steps, N_{rest}. N_{T_I} PITs are used in the acquisition process. Since L is determined from (3.65), let $L_g=N_{max_g}+N_{rest_g}$, so that $N_{max}=\sum_{g=1}^{N_{T_I}}N_{max_g}$ and $N_{rest}=\sum_{g=1}^{N_{T_I}}N_{rest_g}$.

For one cell, in each PIT, N_t values are added. Each value is multiplied by either $+1$ or -1. Since the noise is WGN with zero mean, the PDF of the noise must be an even function. Parts {1} and {2} in (3.67) are normally distributed, so (3.72) satisfies

$$f_{\mathbf{X}}(x)=f_{\mathbf{X}}(-x)$$

The square of the noise amplitude has a distribution as given in (3.74). For the integration in each cell that is calculated by (3.68), the distribution of the maximum $[\mathbf{W}=\max(\mathbf{Y}_1,\mathbf{Y}_2,\ldots,\mathbf{Y}_{(2^{N_t-1})})]$, where $\mathbf{Y}_j$ is the integration result corresponding to a possible data bit combination j, is

$$F_{\mathbf{W}}(w)=F_{\mathbf{Y}_1,\mathbf{Y}_2,\ldots,\mathbf{Y}_{2^{N_t-1}}}(w,\ w,\ \ldots,\ w) \tag{3.84}$$

A bound on that distribution is established as in (3.80) to (3.83). Assuming that all the $\mathbf{Y}_j$'s have the same distribution, defined as $F_{\mathbf{Y}}$, the bound is

$$\bar{F}_{\mathbf{W}}(\gamma) \leq 2^{N_t-1}\,\bar{F}_{\mathbf{Y}}(\gamma) \tag{3.85}$$

This bound is used in the derivation of the p_f only if the accumulation consists of one step. Since the accumulation is calculated over L steps, then the total PDF needs to be derived. For the first $N_{max\,All}$ steps, all the cells apply (3.68). Then, in each of the $N_{maxDelay}$ steps, only N_τ cells apply (3.68). In each of the last N_{rest} steps, only one cell applies (3.68), and the remaining cells apply (3.69). Define $N_{cell_{\tau v}}$ as the number of cells whose powers are the maximum during any of the $N_{maxDelay}$ steps at a code delay τ_v; let $N_{cell_\tau} = \sum_{v=1}^{N_\tau} N_{cell_{\tau v}}$. Define N_{cell_m} as the number of cells whose powers are the maximum during any of the remaining N_{rest} steps. The total PDF for one cell is obtained by convolving L functions, where some of them have a form like $f_{\mathbf{W}}(w)$ derived from (3.84), and the others have a form like $f_{\mathbf{Y}}(y)$ as in (3.74). Such a PDF cannot be obtained in a closed form. Instead, a bound is established on the p_f, or numerical integration and differentiation are used to find the p_f. Both approaches are derived in the following sections.

3.6.4.1 p_f Bound

If the accumulation is determined from (3.68) at a step in a cell, then assume that instead of keeping track of the maximum value only, all 2^{N_t-1} results are kept. This means that each time (3.68) is applied to a cell, the total number of results coming from that cell will be increased by 2^{N_t-1}. On the other hand, if the accumulation is determined from (3.69) at a step in a cell, then the number of results will not increase. For simplicity, assume that $N_{maxDelay} = 0$. Let q_{gj} define the fraction of N_{rest_g} the jth cell, among the N_{cell_m} cells, produced the maximum; q_j defines the fraction of N_{rest} the jth cell produced the maximum. The total number of results N_{Rb} after L steps for an edge b is

$$N_{Rb} = \left[\sum_{j=1}^{N_{cell}-N_{cell_m}} \left[\prod_{g=1}^{N_{T_I}} 2^{\left(N_{t_g}-1\right)N_{max_g}}\right]\right] + \left[\sum_{j=1}^{N_{cell_m}} \left[\prod_{g=1}^{N_{T_I}} 2^{\left(N_{t_g}-1\right)\left(N_{max_g}+q_{gj}\,N_{rest_g}\right)}\right]\right] \tag{3.86}$$

If $N_{maxDelay} \neq 0$, then (3.86) can easily be expanded. Each of the N_b edges will produce a number of results calculated as in (3.86). The total number of results is

$$N_{Result} = \sum_{b=1}^{N_b} N_{Rb} \tag{3.87}$$

Using the distribution in (3.76) and a similar approach as in (3.80) to (3.83), the bound on the p_f is

$$p_f \leq N_{Result} \int_{\gamma}^{\infty} f_{\mathbf{S}}(s)\, ds \tag{3.88}$$

It is difficult to determine N_{cell_m} and q_{gj}, and this approach gives different results for the p_f and the size of data needed for the acquisition when using different values for these variables. For the same data size, the p_f is maximum if $N_{cell_m} = 1$, and p_f decreases as N_{cell_m} increases. However, the difference in the p_f values when assuming different, but close, N_{cell_m} is large. Numerical methods are used to circumvent this problem.

3.6.4.2 p_f with Numerical Methods

To simplify the derivation, from (3.84) assume that the $\mathbf{Y}_j$'s are independent and identically distributed. So,

$$F_{\mathbf{Y}_1} = F_{\mathbf{Y}_2} = \cdots = F_{\mathbf{Y}_{2^{N_{tg}}-1}} \tag{3.89}$$

and

$$F_{\mathbf{W}_g}(w) = F_{\mathbf{Y}_1}(w)\; F_{\mathbf{Y}_2}(w) \;\; \cdots \;\; F_{\mathbf{Y}_{2^{N_{tg}}-1}}(w) = F_{\mathbf{Y}_g}(w)^{2^{N_{tg}}-1} \tag{3.90}$$

where the index g refers to the gth PIT. Assume that the independence assumption will give a probability of false alarm defined as p_{f_u}. Also, assume that if all the $\mathbf{Y}_j$'s are the same, then the probability of false alarm is p_{f_l}. It can be shown that the actual p_f lies between p_{f_u}, as an upper bound, and p_{f_l}, as a lower bound. The $\mathbf{Y}_j$'s are obtained by modulating the integration by the possible data bit combinations. Some combinations can differ in only one bit from some of the others, so the $\mathbf{Y}_j$'s are not independent. As N_{t_g} becomes large, the $\mathbf{Y}_j$'s that result from the data bit combinations differing in only one bit from the correct combination will have values that are close to $\mathbf{Y}_j$ of the correct combination. If

$$\mathbf{Y}_1 = \mathbf{Y}_2 = \cdots = \mathbf{Y}_{2^{N_{tg}}-1} = \mathbf{Y}_g$$

then (3.90) is modified to

$$F_{\mathbf{W}_g}(w) = F_{\mathbf{Y}_g}(w) \tag{3.91}$$

Since $F_{\mathbf{Y}_j}(w) \leq 1$ and $p_f = 1 - F$, then (3.91) will give the smallest p_f. A lower bound on the p_f, defined as p_{fl}, is obtained if all the $\mathbf{Y}_j$'s are equal. This can lead to the conclusion that the actual p_f will satisfy

$$p_{fl} \leq p_f \leq p_{fu} \tag{3.92}$$

The assumption that the $\mathbf{Y}_j$'s are independent will set an upper bound on p_f; thus, it will give a conservative result. This assumption did practically produce acceptable results, as shown in Section 3.7.

The PDF of the integration at one cell, at a step, is obtained by differentiating (3.90). So,

$$f_{\mathbf{W}_g}(w) = 2^{N_{tg}-1} \; f_{\mathbf{Y}_g}(w) \;\; F_{\mathbf{Y}_g}(w)^{2^{N_{tg}-1}-1} \tag{3.93}$$

For N_{maxAll_g} steps, the total PDF, $f_{N_{maxAll_g}}$, for one cell is obtained by convolving N_{maxAll_g} functions as in (3.93). For N_{maxAll} steps, the total PDF, $f_{N_{maxAll}}$, is obtained by convolving N_{T_I} of the $f_{N_{maxAll_g}}$ functions.

In the $N_{maxDelay}$ steps, two different forms of the PDF result from the integration at each cell. If a cell produces the maximum power at a step, then the PDF of the integration at that step has a form similar to (3.93). If a cell does not produce the maximum power at a step, then the PDF of the integration at that step has a form similar to (3.74). The total PDF for the $N_{maxDelay}$ steps, at a cell, is obtained by the convolution of $N_{maxDelay}$ functions with forms such as (3.93) or (3.74); define this PDF as $f_{N_{max\,Delay\epsilon}}$, $\epsilon = 1, \ldots, N_{cell}$.

In the last N_{rest} steps, for the cells that produce the integration using only (3.69), the PDF for one cell has a form like the one in (3.76), but with N_{rest} replacing L. Define this PDF as $f_{N_{rest}}$. The total PDF for each of these cells, after L steps, is determined from

$$f_{\mathbf{S}_{L\epsilon}}(s) = f_{N_{maxAll}} * f_{N_{max\,Delay\epsilon}} * f_{N_{rest}} \tag{3.94}$$

where $*$ defines the convolution operation. For the N_{cell_m} cells, in which throughout the last N_{rest} steps some of their accumulations are calculated from (3.68) and the rest are calculated from (3.69), the total PDF after L steps for the jth cell is determined from

$$f_{S_{L_j}}(s) = f_{N_{maxAll}} * f_{N_{max\,Delayj}} * f_{q_{1\,j}\,N_{rest_1}} * \cdots * f_{q_{N_{T_I}\,j}\,N_{rest_{N_{T_I}}}} * f_{(1-q_j)\,N_{rest}} \tag{3.95}$$

where $f_{q_{gj}N_{rest_g}}$ is a PDF obtained by convolving $q_{gj}N_{rest_g}$ functions of a form like (3.93), while $f_{(1-q_j)N_{rest}}$ is a PDF with a form like that in (3.76), but with $(1-q_j)N_{rest}$ replacing L. The q_{gj} and q_j values are defined as in Section 3.6.4.6.

A bound on the p_f is calculated using a similar approach as in (3.80) to (3.83). So,

$$p_f = \bar{F}_{\mathbf{S}}(\gamma) \leq \sum_{j=1}^{N_{cell_m}} \bar{F}_{\mathbf{S}_{L_j}}(\gamma) + \sum_{\epsilon=1}^{N_{cell}-N_{cell_m}} \bar{F}_{\mathbf{S}_{L_\epsilon}}(\gamma) \tag{3.96}$$

Or, equivalently,

$$p_f \leq \sum_{j=1}^{N_{cell_m}} \left[1 - F_{\mathbf{S}_{L_j}}(\gamma)\right] + \sum_{\epsilon=1}^{N_{cell}-N_{cell_m}} \left[1 - F_{\mathbf{S}_{L_\epsilon}}(\gamma)\right] \tag{3.97}$$

where $F_{\mathbf{S}_{L_j}}(\gamma)$ and $\bar{F}_{\mathbf{S}_{L_j}}(\gamma)$ are obtained by integrating (3.95), respectively, from 0 to γ and from γ to ∞. Similarly, $F_{\mathbf{S}_{L_\epsilon}}(\gamma)$ and $\bar{F}_{\mathbf{S}_{L_\epsilon}}(\gamma)$ are obtained by integrating (3.94). Either (3.96) or (3.97) is used in the p_f calculation, depending on which one is easier to calculate numerically.

3.6.5 Probability of Detection p_d

The probability of detection p_d specifies the probability that (3.67) exceeds γ at the correct cell when a signal is present. The total PDF of the integration calculated from (3.67) is found by first calculating the PDFs that correspond to each of the N_{T_I} PIT intervals and each of the N_{av_g} possible results of the signal averaging. Then, the total PDF is found by convolving all of those PDFs together.

When the PIT is a multiple N_{t_g} of one data bit interval, N_{t_g} coherent integration results are multiplied by the $2^{N_{t_g}-1}$ possible data bit combinations. Then, the N_{t_g} coherent integration results are added together to generate the total coherent integration. This will generate $2^{N_{t_g}-1}$ integration results. But only the integration result that corresponds to the likely data bit combination is kept, and the others are discarded; the likely data bit combination is determined by a maximization operation. Lower C/N_0 has higher BER. This means that there is a possibility that an incorrect data combination will generate higher power as compared to the correct data bit combination. Thus, this possibility must be considered in order to get a correct PDF and, consequently, a correct p_d. Two cases are considered in the following two sections. The first case ignores the effect of the maximization over the possible data bit combination, while the second case considers that effect.

3.6.5.1 p_d Without Considering the Maximization over the Possible Bit Combinations

Consider first the case when the correct data bit combination is chosen, and define G as the number of steps in which that happens. If a signal is present, then a function of the form $r = \sqrt{I^2 + Q^2}$ has the following PDF [5, 7]:

$$f(r) = \frac{r}{\sigma_n^2} e^{-\left(\frac{r^2+A^2}{2\sigma_n^2}\right)} I_0\left(\frac{r\,A}{\sigma_n^2}\right) \tag{3.98}$$

where A is the signal amplitude, σ_n^2 is the noise variance, and I_0 is the modified Bessel function of the first kind, which has the following general form:

$$I_n(x) = \sum_{c=0}^{\infty} \frac{\left(\frac{x}{2}\right)^{2c+n}}{c!\,\Gamma(c+n+1)} \tag{3.99}$$

where $\Gamma(c+n+1) = (c+n)!$

Since (3.67) uses the square of such a function (i.e., $y = r^2$), using a similar approach to that in (3.73) and setting $\sigma_n^2 = 1$, the resultant PDF is

$$f(y) = \frac{1}{2} e^{-\left(\frac{y+A_g^2}{2}\right)} I_0\left(\sqrt{y}\,A_g\right) \tag{3.100}$$

where

$$A_g = \sqrt{2C/N_0\,M_g\,T}$$

$T = 0.001$ sec. Adding G of y will result in a function with a PDF that is the convolution of all of the $f(y)$ functions. The Laplace transform of $f(y)$ is

$$L(f(y)) = \frac{1}{2} e^{-\frac{A_g^2}{2}} \sum_{c=0}^{\infty} \frac{A_g^{2c}}{2^{2c}\;c!\;\left(s+\frac{1}{2}\right)^{c+1}} \tag{3.101}$$

Multiplying a number G of such functions and taking the inverse Laplace transform results in a function of the form

$$f(z) = \frac{1}{2}\left(\frac{z}{A_g^2 G}\right)^{\frac{1}{2}(G-1)} e^{-\left(\frac{z+G\,A_g^2}{2}\right)} I_{G-1}\left(\sqrt{z\,G}\,A_g\right) \tag{3.102}$$

Since each possible data bit combination will give a different average signal power, the value of A_g will depend on that average; define it by

$$A_{gh} = \sqrt{\frac{2C/N_0\,T\,M}{N_{t_g}}}\;(N_{t_g} - 2(h-1)) \tag{3.103}$$

where $h = 1, \ldots, N_{av_g}$ (as discussed at the beginning of Section 3.6.1). Note that $h = 1$ corresponds to the case in which the data bits are estimated correctly. In the case equivalent to "no signal," which is possible only if N_{t_g} is even, $h = N_{t_g}/2 + 1$, or, equivalently, $h = N_{av_g}$. A_{gh} will be reflected in both the I and Q values in (3.62), (3.63), and (3.67).

The summation that results from using each possible combination of L_{gh}, A_{gh}, and T_I generates a PDF of a form similar to (3.102). This is proved as follows. Consider the PDF, f_A, that results from the summation of two functions U_1 and U_2. U_1 and U_2 are generated by summations over a number of functions equal to L_{g1} and L_{g2}, and they generate PDFs of f_1 and f_2, respectively. The L_{g1} and L_{g2} functions have average signal levels of A_{g1} and A_{g2}, respectively. The PDF, f_A, is found by convolving f_1 and f_2 together. The multiplication of the Laplace transform of f_1 and f_2 gives

$$\begin{aligned}
L(f_1)\cdot L(f_2) = \frac{1}{4}\,e^{-\frac{1}{2}\left(L_{g1}A_{g1}^2 + L_{g2}A_{g2}^2\right)} \Bigg[& \frac{1}{2^{L_{g1}+L_{g2}-2}\,(s+\frac{1}{2})^{L_{g1}+L_{g2}}} \\
& + \frac{(A_{g2}\sqrt{L_{g2}})^2}{2^{L_{g1}+L_{g2}}\,1!\,(s+\frac{1}{2})^{L_{g1}+L_{g2}+1}} \\
& + \frac{(A_{g2}\sqrt{L_{g2}})^4}{2^{L_{g1}+L_{g2}+2}\,2!\,(s+\frac{1}{2})^{L_{g1}+L_{g2}+2}} \\
& + \frac{(A_{g2}\sqrt{L_{g2}})^6}{2^{L_{g1}+L_{g2}+4}\,3!\,(s+\frac{1}{2})^{L_{g1}+L_{g2}+3}} \\
& + \frac{(A_{g1}\sqrt{L_{g1}})^2}{2^{L_{g1}+L_{g2}}\,1!\,(s+\frac{1}{2})^{L_{g1}+L_{g2}+1}} \\
& + \frac{(A_{g1}A_{g2}\sqrt{L_{g1}L_{g2}})^2}{2^{L_{g1}+L_{g2}+2}\,1!^2(s+\frac{1}{2})^{L_{g1}+L_{g2}+2}} \\
& + \frac{(A_{g1}\sqrt{L_{g1}})^2(A_{g2}\sqrt{L_{g2}})^4}{2^{L_{g1}+L_{g2}+4}\,1!\,2!\,(s+\frac{1}{2})^{L_{g1}+L_{g2}+3}} \\
& + \frac{(A_{g1}\sqrt{L_{g1}})^2(A_{g2}\sqrt{L_{g2}})^6}{2^{L_{g1}+L_{g2}+6}\,1!\,3!\,(s+\frac{1}{2})^{L_{g1}+L_{g2}+4}}
\end{aligned}$$

$$+\frac{(A_{g1}\sqrt{L_{g1}})^4}{2^{L_{g1}+L_{g2}+2}\,2!\,(s+\frac{1}{2})^{L_{g1}+L_{g2}+2}}$$
$$+\frac{(A_{g1}\sqrt{L_{g1}})^4\,(A_{g2}\sqrt{L_{g2}})^2}{2^{L_{g1}+L_{g2}+4}\,1!\,2!\,(s+\frac{1}{2})^{L_{g1}+L_{g2}+3}}$$
$$+\frac{(A_{g1}\sqrt{L_{g1}})^4\,(A_{g2}\sqrt{L_{g2}})^4}{2^{L_{g1}+L_{g2}+6}\,1!\,3!\,(s+\frac{1}{2})^{L_{g1}+L_{g2}+4}}+\dots\Bigg]$$

The inverse Laplace transform of this function, defined as $f_A(z)$, has the following form:

$$f_A(z)=\frac{1}{4}\,e^{-\left(\frac{z+2\left(L_{g1}A_{g1}^2+L_{g2}A_{g2}^2\right)}{2}\right)}$$
$$\sum_{c=0}^{\infty}\frac{z^{(L_{g1}+L_{g2}+c-1)}\,\left(A_{g1}^2L_{g1}+A_{g2}^2L_{g2}\right)^c}{2^{L_{g1}+L_{g2}+2c-2}\;c!\;\Gamma(L_{g1}+L_{g2}+c)} \tag{3.104}$$

This function is equivalent to

$$f_A(z)=\frac{1}{2}\left(\frac{z}{A_{g1}^2L_{g1}+A_{g2}^2L_{g2}}\right)^{\frac{1}{2}(L_{g1}+L_{g2}-1)}\quad e^{-\frac{1}{2}\left(z+L_{g1}A_{g1}^2+L_{g2}A_{g2}^2\right)}$$
$$I_{L_{g1}+L_{g2}-1}\left(\sqrt{z\,(A_{g1}^2L_{g1}+A_{g2}^2L_{g2})}\right) \tag{3.105}$$

Consider the case in which $A_{g1}=0$; this is equivalent to a "no signal" case. The PDF, f_{A_0}, of the summation of U_1 and U_2 is found by convolving a form like the one in (3.102) with one like (3.76). The multiplication of the Laplace transform of both functions results in

$$\frac{1}{4}\,e^{-\frac{1}{2}\left(L_{g2}A_{g2}^2\right)}\sum_{c=0}^{\infty}\frac{(A_{g2}\sqrt{L_{g2}})^{2c}}{2^{L_{g1}+L_{g2}+2c}\;c!}\;\frac{1}{s^{L_{g1}+L_{g2}+c}} \tag{3.106}$$

The convolution result, defined as $f_{A_0}(z)$, is

$$f_{A_0}(z)=\frac{1}{2}\left(\frac{z}{L_{g2}A_{g2}^2}\right)^{\frac{1}{2}(L_{g1}+L_{g2}-1)}$$
$$e^{-\frac{1}{2}\left(z+L_{g2}A_{g2}^2\right)}\;I_{L_{g1}+L_{g2}-1}(\sqrt{L_{g2}\,z}\;A_{g2}) \tag{3.107}$$

A similar approach can be used to find the PDF, f_T, that results if U_1 and U_2 correspond to different coherent integration lengths. In this case, U_1 and

U_2 are generated by summations over a number of functions equal to L_{1h} and L_{2h}, respectively. The average signal levels of the L_{1h} and L_{2h} functions are A_{1h} and A_{2h}, respectively. The PDF, f_T, has a form similar to the one in (3.105), which is

$$f_T(z) = \frac{1}{2}\left(\frac{z}{A_{1h}^2 L_{1h} + A_{2h}^2 L_{2h}}\right)^{\frac{1}{2}(L_{1h}+L_{2h}-1)} e^{-\frac{1}{2}\left(z+L_{1h}A_{1h}^2+L_{2h}A_{2h}^2\right)}$$
$$\times I_{L_{1h}+L_{2h}-1}\left(\sqrt{z\left(A_{1h}^2 L_{1h} + A_{2h}^2 L_{2h}\right)}\right) \quad (3.108)$$

The PDF of the total integration will be the result of convoluting the PDFs that result from using each possible combination of L_{gh}, A_{gh} and T_I. This PDF is derived by expanding upon the approach used in (3.105), (3.107), and (3.108). In this case, the number of convolved functions is equal to the total number of possible combinations of L_{gh}, A_{gh} and T_I. Define ζ as

$$\zeta = \sum_{g=1}^{N_{T_I}} \sum_{h=1}^{N_{avg}} \left(A_{gh}^2 L_{gh}\right)$$

The total PDF can be shown to have the following form:

$$f(s) = \frac{1}{2}\left(\frac{s}{\zeta}\right)^{\frac{1}{2}(L-1)} e^{-\frac{1}{2}(s+\zeta)} I_{L-1}\left(\sqrt{s\,\zeta}\right) \quad (3.109)$$

So,

$$p_d = \int_{\gamma}^{\infty} f(s)\, ds \quad (3.110)$$

The values of L_{gh}, calculated as in (3.67), depend on the expected C/N_0 and N_{max_g}. When the integration is determined from (3.68) for all the cells, during the N_{maxAll} steps, the probability of choosing the correct data bit combination is set to 1. This is because an incorrect data bit combination will be chosen at the correct cell only if that combination generates higher power as compared to the power generated by the correct data bit combination. Thus, the integration will not degrade. For the remaining N_{rest} steps, there will be a probability of choosing an incorrect data bit combination. When a noise cell produces the maximum value among all the cells, then all of the possible data bit combinations will have equal probabilities of being chosen. The probability that a noise cell will produce the maximum in any of the N_{rest} steps depends on the PDF of the signal and the PDF of the noise at that step. This means that as N_{max} is increased, the probability of choosing an incorrect data bit combination is decreased. Thus, N_{max} should be chosen to be large enough such that, at the end of the N_{max}, steps, it is possible to obtain probabilities of detection and false alarm, defined, respectively, as $p_{d_{Nmin}}$ and $p_{f_{Nmax}}$, that satisfy the following conditions:

$$p_{d_{min}} \leq p_{d_{Nmin}} \leq \bar{p}_d \tag{3.111}$$

$$\bar{p}_f \leq p_{f_{Nmax}} \leq p_{f_{max}} \tag{3.112}$$

where $\bar{p}_f$ and $\bar{p}_d$ are the desired p_f and p_d, respectively. These desired values are given by the user and are used to determine the number of steps, L, needed in the acquisition process. $p_{d_{min}}$ is a probability of detection less than $\bar{p}_d$. $p_{f_{max}}$ is a probability of false alarm greater than $\bar{p}_f$. The probability of choosing an incorrect data bit combination is set depending on $p_{d_{Nmin}}$ and $p_{f_{Nmax}}$.

3.6.5.2 p_d Considering the Maximization over the Possible Bit Combinations

The PDF and the p_d are computed numerically; the following describes how to obtain them.

1. Define $2^{N_{t_g}-1}$ signal levels as A_{nt_g}, for $nt_g = 1, \ldots, 2^{N_{t_g}-1}$. Each signal level corresponds to a possible data bit combination, where

$$A_{nt_g} = \sqrt{\frac{2C/N_0\ T M}{N_{t_g}}}\ \left(N_{t_g} - 2\, e_{nt_g}\right) \tag{3.113}$$

   where e_{nt_g} is the number of data bits, within the N_{t_g} data bits, with the wrong signs.

2. Define F_{nt_g} as the distribution that results from the integration, at one acquisition step, if the signal level is A_{nt_g}. Define f_{nt_g} as the PDF that results from the integration if the signal level is A_{nt_g}; this PDF has a form similar to (3.100).

3. The distribution that results from the maximization over the $2^{N_{t_g}-1}$ integrations, at the lth step of the acquisition, is approximated as

$$F_{gl} = F_{1_g}\ F_{2_g} \ldots F_{2_g^{N_{tg}-1}} \tag{3.114}$$

4. The PDF that corresponds to the distribution in (3.114) is found by differentiating (3.114). So,

$$f_{gl} = f_{1_g}\ F_{2_g} \ldots F_{2_g^{N_{tg}-1}} + f_{2_g}\ F_{1_g}\ F_{3_g} \ldots F_{2_g^{N_{tg}-1}} + \ldots$$
$$+ F_{1_g}\ F_{2_g} \ldots f_{2_g^{N_{tg}-1}} \tag{3.115}$$

5. Since there is a possibility of using N_{T_I} PIT lengths, and each PIT is used for L_g acquisition steps, the total PDF that results from each PIT is found by convolving L_g functions of a form as in (3.115). Define this PDF as f_g. The total PDF that results after L acquisition steps is found by convolving the N_{T_I} functions of f_g. Define the total PDF as $f_{N_{T_I}}$. Thus, the p_d is

$$p_d = \int_{\gamma}^{\infty} f_{N_{T_I}}(s)\, ds \tag{3.116}$$

If the general approach of choosing the likely data bit combination is followed, then there is a possibility that a noise cell could produce the maximum power. In this case, the integration in the correct cell is not chosen by a maximization process. The PDF of the integration at the correct cell, at the acquisition steps that have a noise cell producing the maximum power, will have a form as in (3.100). Thus, when calculating the total PDF, the PDF f_g is replaced with the PDF in (3.100) at those steps.

3.6.6 Threshold Calculation

The threshold γ is found by a search process. Every γ and L, the total number of acquisition steps, will give a certain p_f and p_d. The desired p_f and p_d are $\bar{p}_f$ and $\bar{p}_d$, respectively. The search is performed to find γ and L that produce $p_f(\gamma, L)$ and $p_d(\gamma, L)$ that satisfy the following conditions:

$$p_f(\gamma, L) \le \bar{p}_f$$

$$p_d(\gamma, L) \ge \bar{p}_d$$

3.6.7 Signal Detection with Loss

Three factors affect the probability of detection:

(1) *Doppler separation.* This causes a maximum loss if the true Doppler shift lies between two Doppler bins. This loss causes a reduction in the coherent integration by a value equal to $\text{sinc}(f_e\, T_I)$. f_e is the difference between the Doppler bin and the true Doppler shift.

(2) *Code-delay error.* This causes a maximum loss if the true code delay lies between two samples. This loss causes the autocorrelation function $R(\tau_e)$ to be equal to

$$R(\tau_e) = 1 - \frac{\tau_e}{T_{chip}}$$

where τ_e is the code-delay error. T_{chip} is the chip length.

(3) *Loss because the correct bit edge is not one of the chosen N_b edges.* This is modeled as a reduction in the data bit length as

$$M_{loss} = M - 2\left\lfloor \frac{M}{2N_b} \right\rfloor p_r$$

where p_r is the probability of a data transition.

Such losses cause (3.103) to become

$$A_{gh} = \mathrm{R}(\tau_e)\ \mathrm{sinc}(f_e\ T_I)\sqrt{\frac{2C/N_0\ T\ M_{loss}}{N_{t_g}}}\ (N_{t_g} - 2(h-1)) \tag{3.117}$$

In this case, p_d is calculated using (3.117) to account for the losses.

3.6.8 p_d and p_f for Signal Detection in the Presence of Strong Interfering Signals

The derivations of p_d, p_f and γ for the algorithm that detects weak signals in the presence of strong interfering signals follows directly from the derivations made in the previous sections. In that algorithm, to be able to use a PIT multiple N_t of T_{dms}, the following settings are used: $N_{max} = L$, $N_{rest} = 0$, and $N_{cell_m} = N_{cell}$. The p_d and p_f are calculated taking these settings into consideration. The p_f is calculated numerically as described in the previous sections.

Also, the calculation of the probability of detection of an interfering signal p_{DC} is calculated by a method similar to the one used to calculate p_d, except that in this case, the signal level is set equal to

$$A_{C_{gh}} = signalReduction\ \ A_{gh} \tag{3.118}$$

Where, *signalReduction* is about 0.15 in the worst case (i.e., the case that produces the maximum interfering power); otherwise, it is smaller.

3.7 Simulation and Results

The developed algorithms are demonstrated using simulated GPS signals with C/A codes and TCXO clocks. The codes are modulated by a ± 1 data with a rate of 50 Hz and a probability of data transition equal to 0.5. The following values are used: The IF frequency $f_{IF} = 1{,}405$ kHz. $f_{L1} = 1{,}575.42$ MHz. The initial phase is modeled as a uniformly distributed RV between $(-\pi,\ \pi)$ radians. A Doppler shift range between $(-5,\ 5)$ kHz is assumed. The sampling rate is set to $f_s = 3{,}050$ kHz for the CCMDB algorithm, and it is set to $f_s = 5{,}700$ KHz for the MDBZP algorithm. The oscillator phase and frequency noises are modeled with normal random walks; the model is similar to the one in [8], and the variances of the random walks are derived as in [9]. A TCXO is simulated. The values of the phase and frequency random-walk intensities S_f and S_g are similar to those used in [10], where $S_f = 5 \times 10^{-21}$ s, $S_g = 5.9 \times 10^{-20}$ s^{-1}.

The performances of the two new acquisition algorithms presented in this chapter (i.e., CCMDB and MDBZP) are compared with two previous acquisition algorithms that were presented in [3, 11]. The first previous acquisition algorithm was developed in [3], and it uses a PIT of 20 ms. The second previous acquisition algorithm was developed in [11], and it uses PIT of 10 ms. In the simulation of the algorithm of [3], with the PIT of 20 ms, only the accumulation that starts at the correct bit edge position is used in the test. In the simulation of the algorithm of [11], where the PIT is 10 ms, only the half of the data bit interval that does not have a data transition is used in the test. To present a fair comparison between different algorithms, the same amount of received data is used in the test of each algorithm.

The first algorithm (CCMDB) results are as follows. Values of C/N_0 of 17, 15, and 10 dB-Hz are tested using $N_b = 4$. The four edges are located at offsets of 2, 7, 12, and 17 ms relative to the start of the data bits, which is the worst case if $N_b = 4$. Six different PITs are used in the test of the developed algorithm. The PIT starts at 20 ms and gradually increases by 20 ms. In all cases, the true code delay was 307.2 chips, and the Doppler shift was 2 kHz. Table 3.1 summarizes the settings used with each C/N_0, and Table 3.2 summarizes the acquisition results.

For the $C/N_0 =$ 17 dB-Hz case: Each PIT, from 20 ms to 120 ms, runs for a number of steps equal to 10, 10, 5, 2, 2, and 1, with a total of 30 steps. The Doppler bin elimination is done every 5 steps until step 25; then, it is done every 2 steps. The Doppler elimination variables are set as follows: $E_{fz} = [0.3\ \ 0.35\ \ 0.4\ \ 0.5\ \ 0.7\ \ 0.8]$; $E_{change} = [-0.05\ \ -0.05\ \ 0\ \ 0\ \ 0]$; $E_{limit} = [0.2\ \ 0.3\ \ 0.4\ \ 0.5\ \ 0.7\ \ 0.8]$; $N_{d_{min}} = [80\ \ 40\ \ 40\ \ 10\ \ 5\ \ 5]$. $N_{max} = 29$ ($N_{maxAll} =$ 15 and $N_{maxDelay} = 14$); $N_{cell_m} = 2$. The analysis of this test case results in $p_f = 10^{-5}$, $p_d = 0.9$, and $\gamma = 170$. The simulation gave the following results: The estimated Doppler shift was 2.00125 kHz. The maximum power in the noise cells at the last step had a value of 152. The correct cell had a power of 222. Figure 3.9(a) shows the acquisition result of the developed algorithm. Figure 3.9(b) shows the acquisition result using a PIT of 20 ms and 69 steps; there was no positive acquisition as the correct cell did not produce the maximum power. Figure 3.9(c) shows the acquisition result using a PIT of 10 ms using 138 steps. There was no positive acquisition; the correct cell did not produce the maximum power.

For the $C/N_0 =$ 15 dB-Hz case: Each PIT, from 20 ms to 120 ms, runs for a number of steps equal to 30, 30, 4, 4, 4, and 7, with a total of 79 steps. The Doppler bin elimination is done every 15 steps until step 60; then, it is done every 4 steps. The Doppler elimination variables are set as follows: $E_{fz} = [0.2\ \ 0.25\ \ 0.6\ \ 0.7\ \ 0.7\ \ 0.8]$; $E_{change} = [-0.05\ \ \ 0.05\ \ 0\ \ 0\ \ 0]$; $E_{limit} = [0.15$

Table 3.1
CCMDB Acquisition Algorithm Settings

Parameter	17 dB-Hz	15 dB-Hz	10 dB-Hz
Initial PIT length (ms)	20	20	20
Final PIT length (ms)	120	120	120
# steps for each PIT	[10; 10; 5; 2; 2; 1]	[30; 30; 4; 4; 4; 7]	[60; 45; 20; 20; 10; 16]
Total number of steps	30	79	171
First N_{ez}	5	15	15
Change N_{ez} after	25 steps	60 steps	105 steps
Second N_{ez}	2	4	10
E_{fz}	[0.3; 0.35; 0.4; 0.5; 0.7; 0.8]	[0.2; 0.25; 0.6; 0.7; 0.7; 0.8]	[0.35; 0.2; 0.25; 0.4; 0.45; 0.6]
E_{change}	[−0.05; −0.05; 0; 0; 0]	[−0.05; 0.05; 0; 0; 0]	[−0.25; 0.05; 0; 0; 0]
E_{limit}	[0.2; 0.3; 0.4; 0.5; 0.7; 0.8]	[0.15; 0.3; 0.6; 0.7; 0.7; 0.8]	[0.05; 0.25; 0.25; 0.4; 0.45; 0.6]
$N_{d_{min}}$	[80; 40; 40; 10; 5; 5]	[80; 40; 40; 10; 5; 5]	[100; 80; 80; 60; 20; 5]
N_{max}	29	70	160
$N_{max\,All}$	15	40	160
$N_{max\,Delay}$	14	30	0
N_{cell_m}	2	2	2

Table 3.2
The Acquisition Results of the CCMDB Algorithm

Parameter	17 dB-Hz	15 dB-Hz	10 dB-Hz
p_f	10^{-5}	10^{-5}	.004
p_d	0.9	0.97	0.9
γ	170	320	715
Signal power (last step)	222	371	745
Maximum noise power (last step)	152	307	640
Estimated f_d KHz	2.00125	2.00125	2.005

0.3 0.6 0.7 0.7 0.8]; $N_{d_{min}} = [80\ \ 40\ \ 40\ \ 10\ \ 5\ \ 5]$; $N_{max} = 70$ ($N_{maxAll} =$ 40 and $N_{maxDelay} = 30$); $N_{cell_m} = 2$. The analysis of this test case results in $p_f = 10^{-5}$, $p_d = 0.97$, and $\gamma = 320$. The simulation gave the following results: The estimated Doppler shift was 2.00125 kHz. The maximum power in the noise cells at the last step had a value of 307. The correct cell had a power of 371. Figure 3.10(a) shows the acquisition result of the developed algorithm. Figure 3.10(b) shows the acquisition result using PIT of 20 ms and 180 steps; there was no positive acquisition. The 10-ms acquisition method was tested using 360 steps. There was no positive acquisition; the correct cell did not produce the maximum power.

For the $C/N_0 =$ 10 dB-Hz case: Each PIT, from 20 ms to 120 ms, runs for a number of steps equal to 60, 45, 20, 20, 10, and 16, with a total of 171 steps. The Doppler bin elimination is done every 15 steps until step 105; then, it is done every 10 steps. The Doppler elimination variables are set as follows: $E_{fz} = [0.35\ \ 0.2\ \ 0.25\ \ 0.4\ \ 0.45\ \ 0.6]$; $E_{change} = [-0.25\ \ 0.05\ \ 0\ \ 0\ \ 0]$; $E_{limit} = [0.05\ \ 0.25\ \ 0.25\ \ 0.4\ \ 0.45\ \ 0.6]$; $N_{d_{min}} = [100\ \ 80\ \ 80\ \ \ 60\ \ 20\ \ 5]$; $N_{max} = 160$ and $N_{cell_m} = 2$. The analysis of this test case results in $p_f = 0.004$, $p_d = 0.9$, and $\gamma = 715$. The simulation gave the following results: The estimated Doppler shift was 2.005 kHz. The maximum power in the noise cells at the last step had a value of 640. The correct cell had a power of 745. Figure 3.11 shows the acquisition result of the developed algorithm. The acquisition test using PIT of 20 ms and 436 steps has failed as the correct cell did not produce the maximum power. The 10-ms acquisition method also failed the acquisition test using 872 steps.

The second developed acquisition algorithm, MDBZP, results are as follows: Values of C/N_0 of 17, 15, and 10 dB-Hz are tested using $N_b = 10$. The 10 edges are located at offsets of 1, 3, 5, . . . , 19 ms relative to the start of the data bits, which is the worst case if $N_b = 10$. A PIT of 80 ms is used in the test of the developed algorithm. In all cases, the true code delay was 307.4 chips, and the Doppler shift was 0.5 kHz. Table 3.3 summarizes the acquisition result of the MDBZP.

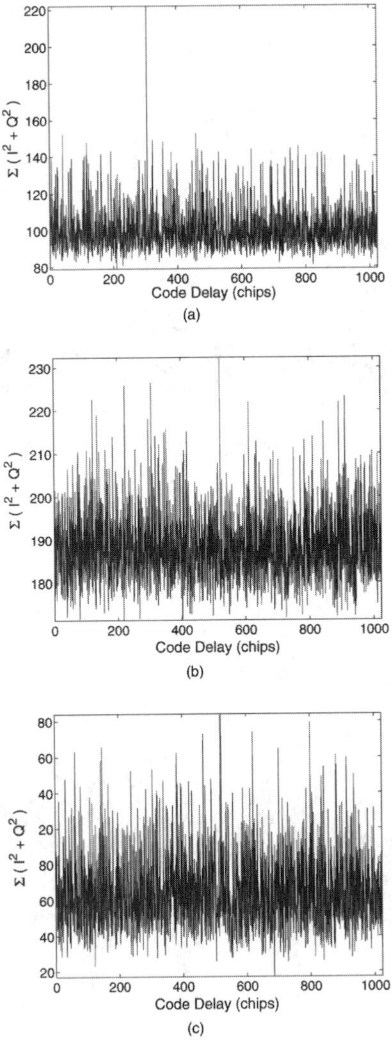

Figure 3.9 Incoherent accumulation result (power) versus code delay of the acquisition test of a 17-dB-Hz signal: (a) using the CCMDP algorithm with 30 steps, starting at an offset of 2 ms relative to the correct bit edge position; (b) using a PIT of 20 ms with 69 steps, starting at the correct bit edge position; the test gave a failed acquisition; and (c) using a PIT of 10 ms with 138 steps; the test gave a failed acquisition.

For the $C/N_0 = 17$ dB-Hz case: The acquisition runs for 18 steps, with $N_{max} = 12$ ($N_{maxAll} = 7$ and $N_{maxDelay} = 5$), and $N_{cell_m} = 2$. The analysis of this test case results in $p_f = 10^{-5}$, $p_d = 0.99$, and $\gamma = 150$. The simulation gave the following results: The maximum power in the noise cells at the last step had a value of 131. The correct cell had a power of 182. Figure 3.12(a) shows the

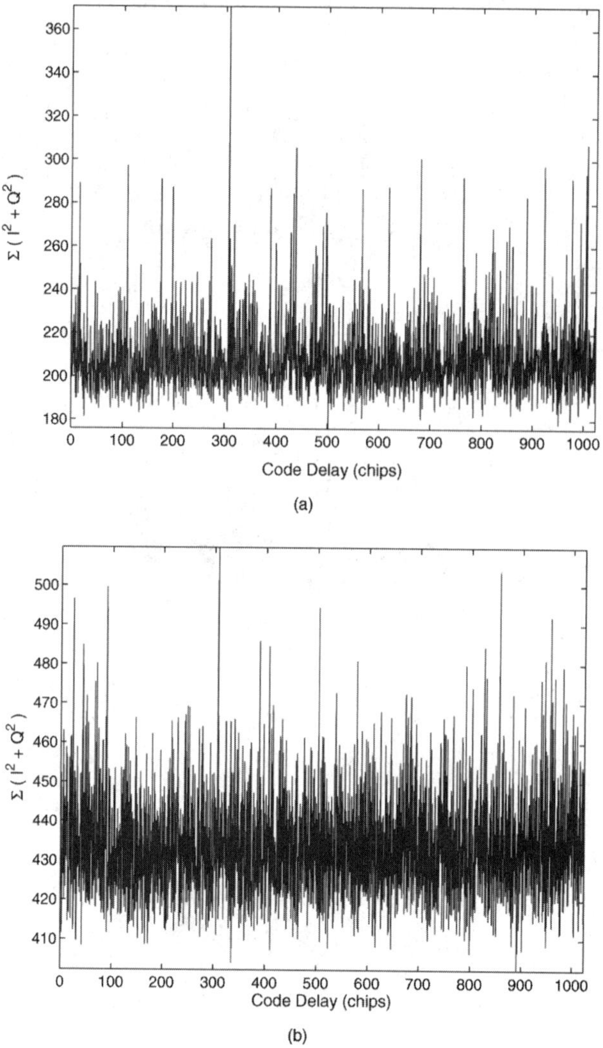

Figure 3.10 Incoherent accumulation result (power) versus code delay of the acquisition test of a 15-dB-Hz signal: (a) using the CCMDP algorithm with 79 steps, starting at an offset of 2 ms relative to the correct bit edge position; and (b) using a PIT of 20 ms with 180 steps, starting at the correct bit edge position; the test gave a failed acquisition.

acquisition result of the developed algorithm. Figure 3.12(b) shows the acquisition result using a PIT of 20 ms and 72 steps. There was no positive acquisition as the correct cell did not produce the maximum power. Figure 3.12(c) shows the acquisition result using a PIT of 10 ms and 144 steps. There was no positive acquisition as the correct cell did not produce the maximum power.

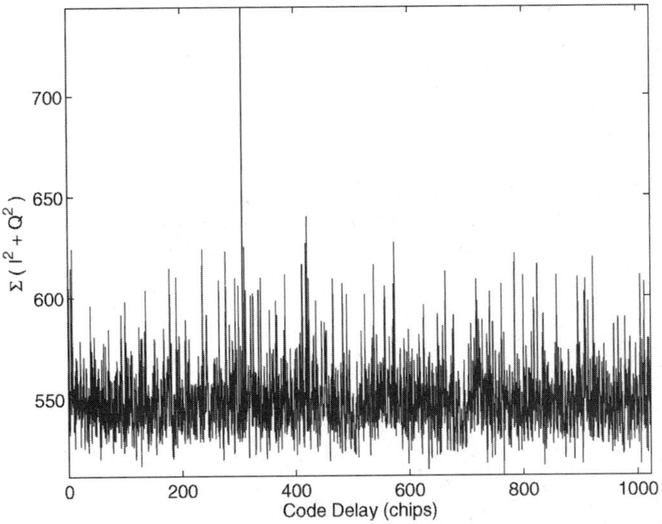

Figure 3.11 Incoherent accumulation result (power) versus code delay of the acquisition test of a 10-dB-Hz signal using the CCMDP algorithm with 171 steps, starting at an offset of 2 ms relative to the correct bit edge position.

An important parameter in the MDBZP algorithm is N_{range}. If the correct Doppler shift is not equal to one of the f_{mid_i} frequency values, then there will be some loss depending on the difference between the correct Doppler shift and the nearest f_{mid_i}. Three different cases are tested in which the correct Doppler shift is located at different distances from the nearest f_{mid_i}. The three distances are 0, 0.5 kHz, and 1 kHz. In all the three tests, the same noise set is used to give accurate indication about the performance difference. Figure 3.12(a) shows the result when the Doppler shift is equal to one of the f_{mid_i}; the power in the correct cell had a value of 182. Figure 3.13(a) shows the result when the Doppler shift is separated from the nearest f_{mid_i} by 0.5 kHz; the power in the correct cell

Table 3.3
The Acquisition Results of the MDBZP Algorithm

Parameter	**17 dB-Hz**	**15 dB-Hz**	**10 dB-Hz**
p_f	10^{-5}	10^{-5}	10^{-4}
p_d	0.99	0.9	0.9
γ	150	200	570
Signal power (last step)	182	270	521
Maximum noise power (last step)	131	220	628

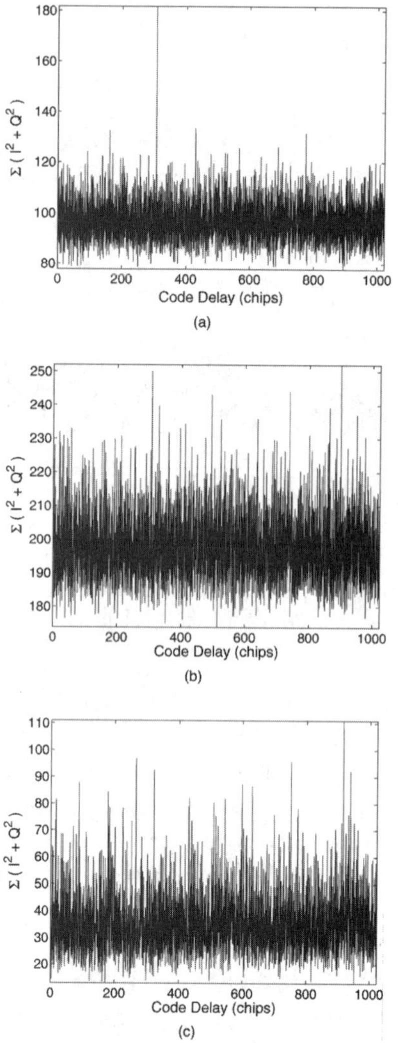

Figure 3.12 Incoherent accumulation result (power) versus code delay of the acquisition test of 17-dB-Hz signal: (a) using the MDBZP algorithm with 18 steps and a PIT of 80 ms, starting at an offset of 1 ms relative to the correct bit edge position; (b) using a PIT of 20 ms with 72 steps, starting at the correct bit edge position; the test gave a failed acquisition; and (c) using a PIT of 10 ms with 144 steps; the test gave a failed acquisition.

had a value of 177, which represents about 3% power loss. Figure 3.13(b) shows the result when the Doppler shift is separated from the nearest f_{mid_i} by 1 kHz; the power in the correct cell had a value of 169.5, which represents about 7% power loss. Thus, a range in size of 2 kHz can be used since the worst case of correct Doppler shift location will be at 1 kHz from the nearest f_{mid_i}. In addition,

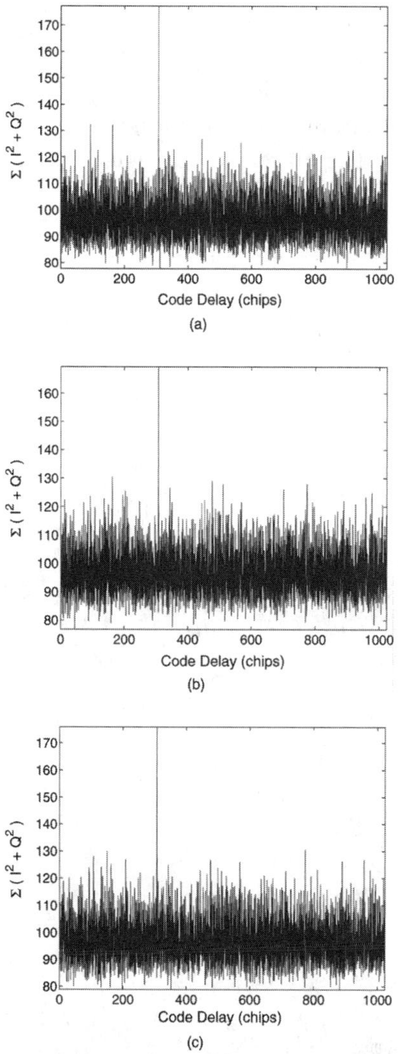

Figure 3.13 Incoherent accumulation result (power) versus code delay of the acquisition test of a 17-dB-Hz signal using the MDBZP algorithm: (a) the correct Doppler shift is separated from the nearest f_{mid_i} by 0.5 kHz; (b) The correct Doppler shift is separated from the nearest f_{mid_i} by 1 kHz; (c) the correct Doppler shift is separated from the nearest f_{mid_i} by 1 kHz, the start of the code is located in the middle of two samples, and the correct Doppler shift is equal to 0.5–$(1/(2T_I))$ kHz.

the worst case of code-delay location is tested in which the start of the code is located in the middle of two samples, and with the correct Doppler shift equal to $0.5 - (1/(2T_I))$ kHz, where $T_I = 80$ ms. Figure 3.13(c) shows this result in which the correct cell had a power of 176.

For the $C/N_0 =$ 15 dB-Hz case: The acquisition runs for 30 steps, with $N_{max} = 20$ ($N_{maxAll} = 10$ and $N_{maxDelay} = 10$), and $N_{cell_m} = 2$. The analysis of this test case results in $p_f = 10^{-5}$, $p_d = 0.9$, and $\gamma = 220$. The simulation gave the following results: The maximum power in the noise cells at the last step had a value of 200. The correct cell had a power of 270. Figure 3.14(a) shows the acquisition

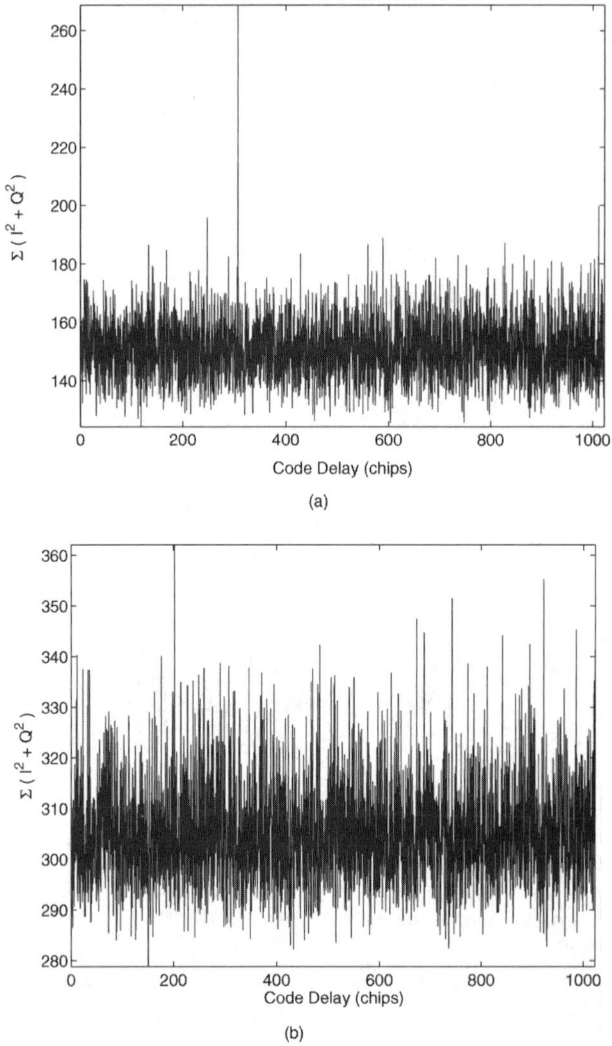

Figure 3.14 Incoherent accumulation result (power) versus code delay of the acquisition test of a 15-dB-Hz signal: (a) using the MDBZP algorithm with 30 steps and a PIT of 80 ms, starting at an offset of 1 ms relative to the correct bit edge position; and (b) using a PIT of 20 ms with 130 steps, starting at the correct bit edge position; the test gave a failed acquisition.

result of the developed algorithm. Figure 3.14(b) shows the acquisition result using a PIT of 20 ms and 120 steps; there was no positive acquisition. The 10-ms acquisition method was tested using 240 steps. There was no positive acquisition as the correct cell did not produce the maximum power.

For the $C/N_0 =$ *10 dB-Hz case*: The acquisition runs for 104 steps, with $N_{max} = 70$ ($N_{maxAll} = 10$ and $N_{maxDelay} = 60$), and $N_{cell_m} = 2$. The analysis of this test case results in $p_f = 10^{-4}$, $p_d = 0.9$, and $\gamma = 570$. The simulation gave the following results: The maximum power in the noise cells at the last step had a value of 521. The correct cell had a power of 628. Figure 3.15 shows the acquisition result of the developed algorithm. The acquisition test using a PIT of 20 ms and 416 steps failed as the correct cell did not produce the maximum power. The 10-ms acquisition method also failed the acquisition test using 832 steps.

The detection in the presence of strong interfering signals is tested. The received signal contained three different satellite signals, with PRN 18, 22, and 3. The three signals had C/N_0 of 20, 35, and 45 dB-Hz, respectively, and they had Doppler shifts of 0.25, 0 and 0.5 kHz, respectively. The acquisition of the 20-dB-Hz signal using $T_I = 40$ ms, $p_f = 0.001$, and $p_{DC} = 0.1$ requires a number of steps equal to 75 and $\gamma = 630$. Figure 3.16 shows the result. To test the acquisition, each Doppler bin that crosses the threshold is inspected at all of its code delays. Each Doppler bin that crosses the threshold at many code delays is

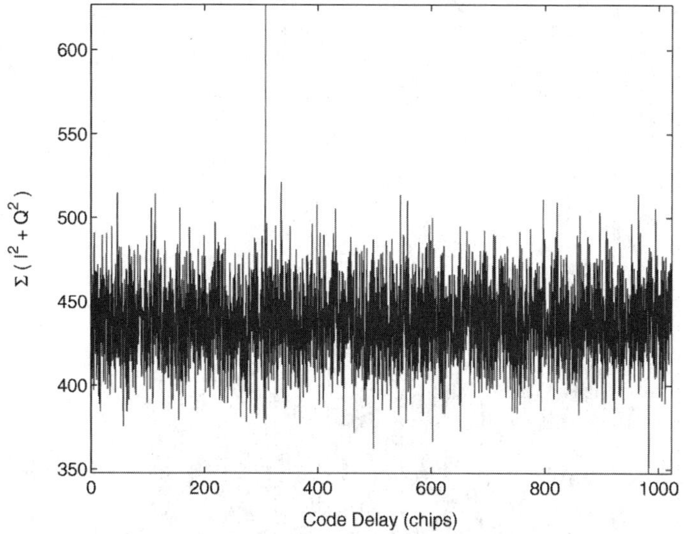

Figure 3.15 Incoherent accumulation result (power) versus code delay of the acquisition test of a 10-dB-Hz signal using the MDBZP algorithm with 104 steps and a PIT of 80 ms, starting at an offset of 1 ms relative to the correct bit edge position.

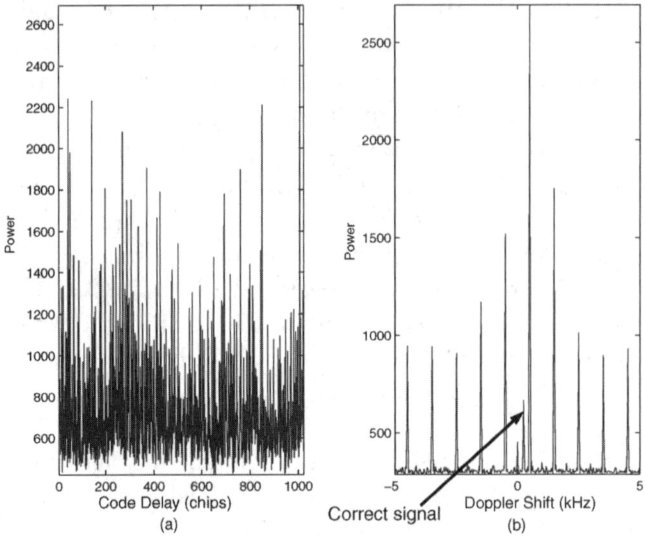

Figure 3.16 Acquisition result of a 20-dB-Hz signal with interfering signals of 35 and 45-dB-Hz; (a) power versus code delay, and (b) power versus Doppler shift.

discarded. Only the correct signal passed this test, where the power reached a value of about 665 at the correct code delay. Figure 3.17 shows the power versus the code delays at the Doppler shift of 0.5 kHz, which is the Doppler shift of the 45-dB-Hz interfering signal. At this Doppler shift, many code delays exceeded the threshold,

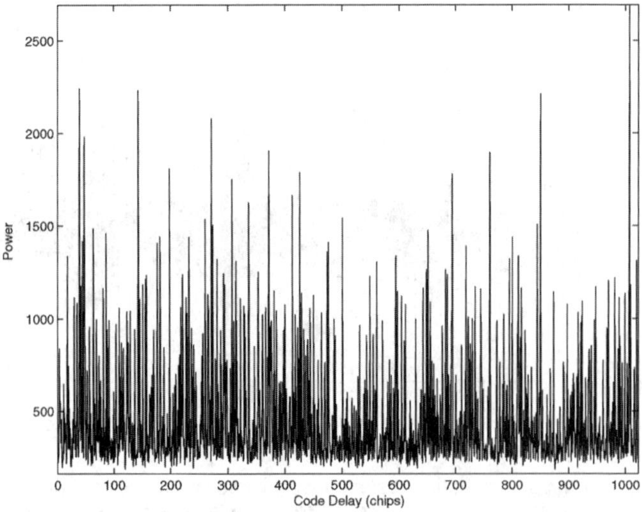

Figure 3.17 Power versus code delay at the Doppler shift of the 45-dB-Hz interfering signal.

so it is discarded. Figure 3.18 shows the power versus the code delays at the Doppler shift of 0.25 kHz (i.e., the 20-dB-Hz acquired signal); only one code delay exceeds the threshold. This test enabled the acquisition of the 20-dB-Hz signal.

Also, high dynamic acquisition is tested. The acquisition of a 20-dB-Hz signal with a Doppler rate of 200 Hz/s is illustrated in Figures 3.19 and 3.20.

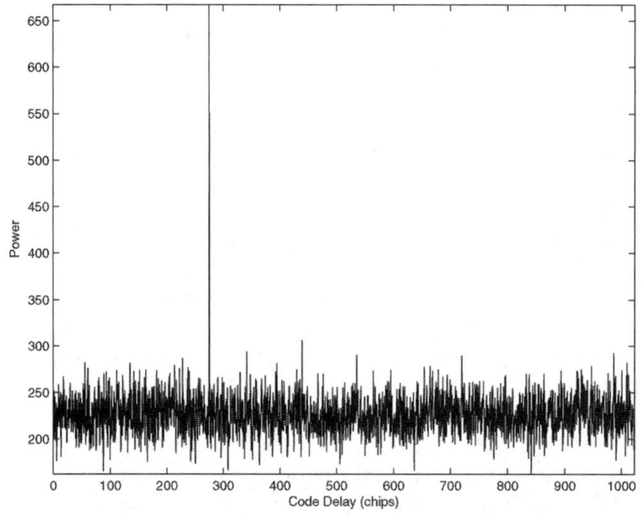

Figure 3.18 Power versus code delay at the correct Doppler shift for the acquired 20-dB-Hz signal.

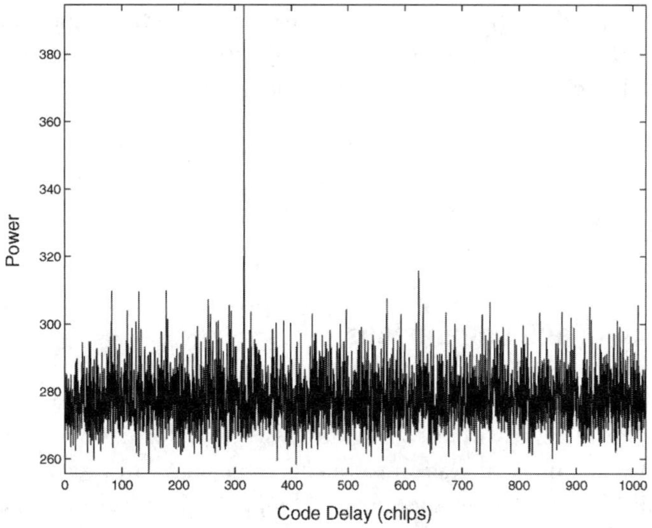

Figure 3.19 Power versus code delay of the acquisition of a 20-dB-Hz signal, with a Doppler rate of 200 Hz/s.

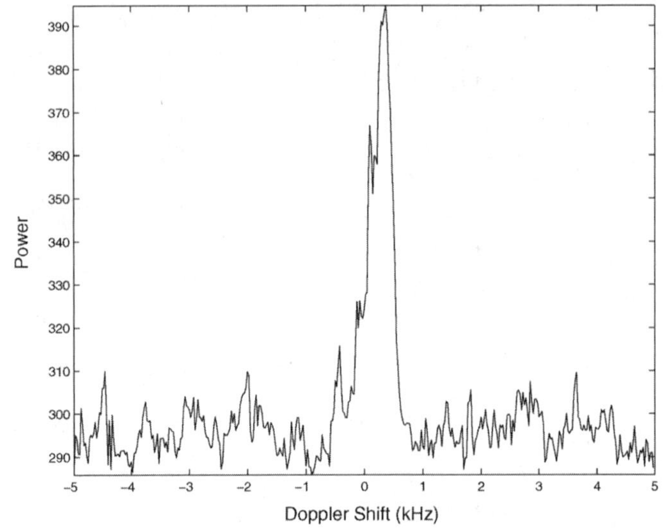

Figure 3.20 Power versus Doppler shift of the acquisition of a 20-dB-Hz signal, with a Doppler rate of 200 Hz/s.

Figure 3.19 shows the power versus the code delays, while Figure 3.20 shows the power versus Doppler shift. This result is obtained using 40-ms coherent integration and a total of 1.8 sec. of data. S_m is equal to one step. The starting Doppler shift was 6.25 Hz, so it was about 366.25 Hz at the end of the acquisition. The Doppler shift that had the maximum power at the end of the acquisition process was 375 Hz, and the estimated Doppler rate was 188 Hz/s. Since the acquired Doppler shift is the average Doppler shift over the last coherent integration interval, then the estimated Doppler shift is calculated as

$$\begin{aligned}\hat{f}_d &= 375 - 188\,\frac{40}{(2)(1,000)} \\ &= 371 \text{ Hz}\end{aligned}$$

3.8 Summary and Conclusions

Two weak-signal acquisition algorithms are developed to use a PIT multiple of one data bit interval without using any assisting information. The CCMDB algorithm uses circular correlation to calculate the coherent integration. It starts with a small PIT, eliminates the unlikely Doppler bins, and then increases the PIT. The second algorithm uses MDBZP to calculate the coherent integration; it circumvents the DBZP limitations resulting from the unaccounted for effect of Doppler shift on the code length. Also, algorithms are developed to detect weak signals in the presence of strong interfering signals and to detect weak signals under high dynamic conditions.

This chapter also provides analysis and derivation for the probabilities of false alarm, p_f, and detection, p_d, and the threshold calculation for the developed algorithms. The analysis takes into consideration all of the approaches used in those algorithms. A bound on the p_f is established to account for the method used to estimate the most likely data bit combination and to account for the total number of searched cells after the Doppler bin elimination. Also, a method to calculate the p_f numerically is introduced. The PDF of the integration result in the correct cell is derived accurately, and then it is used to calculate the p_d. The p_f and p_d are used to calculate the acquisition threshold, γ. The analysis can easily be extended and used in any acquisition algorithm.

The results indicate the ability of the algorithms to detect signals as low as 10 dB-Hz and signals with Doppler rates higher than 200 Hz/s. Large PIT lengths are used with reduced processing and storage requirements.

References

[1] Cormen, T., C. Leisersen, and R. Rivest, *Introduction to Algorithms*, Boston: MIT Press, 1990.

[2] Mattos, P. G., "Solutions to the Cross-Correlation and Oscillator Stability Problems for Indoor C/A Code GPS," *Proc. ION GPS*, Portland, OR, September 9–12, 2003, pp. 654–659.

[3] Psiaki, M. L., "Block Acquisition of Weak GPS Signals in a Software Receiver," *Proc. ION GPS*, Salt Lake City, Utah, September 11–14, 2001, pp. 2838–2850.

[4] Madhani, P. H., et al., "Application of Successive Interference Cancellation to the GPS Pseudolite Near-Far Problem," *IEEE Trans on Aerospace and Electronic Systems*, Vol. 39, No. 2, April 2003, pp. 481–488.

[5] Papoulis, A., *Probability, Random Variables, and Stochastic Processes*, 3rd ed., New York: WCB McGraw-Hill, 1991.

[6] Arnold, B. C., and N. Balakrishnan, *Relations, Bounds and Approximations for Order Statistics*, New York: Springer-Verlag, 1989.

[7] Van Trees, H. L., *Detection, Estimation, and Modulation Theory*, New York: John Wiley & Sons, 1968.

[8] Brown, R. G., and P. Y. C. Hwang, *Introduction to Random Signals and Applied Kalman Filtering*, New York: John Wiley & Sons, 1992.

[9] Van Dierendonck, A. J., J. B. McGraw, and R. G. Brown, "Relationship between Allan Variances and Kalman Filter Parameters," *Proc. 16th PTTI Application and Planning Meeting, NASA Goddard Space Flight Center*, November 27–29, 1984, Greenbelt, MD, pp. 273–293.

[10] Psiaki, M. L., and H. Jung. "Extended Kalman Filter Methods for Tracking Weak GPS Signals," *Proc. ION GPS*, Portland, OR, September 24–27, 2002, pp. 2539–2553.

[11] Lin, D. M., and J. B. Y. Tsui. "A Software GPS Receiver for Weak Signals, *Proc. IEEE MTT-S Digest*, 2001, Phoenix, AZ, May 20–25, 2001, pp. 2139–2142.

4

Fine Acquisition, Bit Synchronization, and Data Detection

4.1 Introduction

A bit synchronization and data bit detection algorithm is developed in this chapter. The algorithm is based on the Viterbi algorithm (VA), and it is used as a basis for two other algorithms. The first algorithm is a fine acquisition algorithm, defined as extended states VA (ES-VA). It provides more accurate estimation for the carrier parameters (i.e., the Doppler shift, Doppler rate, and phase), following a coarse acquisition as discussed in Chapter 3, Sections 3.2 and 3.3. The second algorithm is a bit synchronization and navigation message detection algorithm. It uses extended Kalman filters (EKFs) in the VA paths to track and remove the errors in the carrier parameters. These algorithms are developed to be able to use the output of any acquisition algorithm, not only those developed in Chapter 3.

The VA is a technique based on dynamic programming [1]. It estimates the most likely sequence of data given a set of 1-ms complex correlations. This operation is done through an optimal recursive search process. Each possible bit edge position is represented by a separate state diagram. A weight function, derived from a log-likelihood function of the received signal value, is assigned to each transition in the trellis graph. The path that generates the minimum cumulative weight of its transitions contains the estimated data sequence and corresponds to the estimated bit edge position.

The bit synchronization algorithm is compared to the conventional histogram method [2]. The result shows a superior performance of the bit synchronization algorithm at very weak signal powers. The bit synchronization achieves a high bit edge detection rate (EDR) and an optimal bit error rate (BER) even at very-low-power signals (15 dB-Hz). The BER is reduced further by utilizing the fact that some subframes are repeated every 30 seconds.

This chapter is organized as follows. Section 4.2 presents an overview of the signal model. Section 4.3 presents an overview of the relationship between the acquisition, fine acquisition, bit synchronization, and signal generator modules. Section 4.4 presents a description of the basis algorithm for bit synchronization and data sequence estimation. Section 4.5 presents the ES-VA for fine acquisition. Section 4.6 presents high dynamic fine acquisition. Section 4.7 expands the bit synchronization and data sequence estimation basis algorithm and integrates it with an EKF; this is to allow its usage when there are carrier-tracking errors. Section 4.8 presents a method that utilizes the repeated subframes of the navigation message to reduce the minimum BER that can be achieved. Section 4.9 provides a computational analysis of the developed algorithms. Section 4.10 presents the simulation and results. Section 4.11 presents the summary and conclusion.

4.2 Signal Model

After an acquisition process, an approximate Doppler shift $\hat{f}_d$, Doppler rate $\hat{\alpha}$, phase $\hat{\theta}$, and code delay $\hat{\tau}$ are obtained. The model for the correlated signals over one code period, T_{c_i}, at the ith interval, which follows directly from (2.31) and (2.32), is

$$I_i = A\, d_i\, R(\tau_{e_i})\ \mathrm{sinc}\left(\left(f_{e_i} + \alpha_e \frac{T_{c_i}}{2}\right) T_{c_i}\right) \cos(\theta_{e_i}) + n_{I_i} \tag{4.1}$$

$$Q_i = A\, d_i\, R(\tau_{e_i})\ \mathrm{sinc}\left(\left(f_{e_i} + \alpha_e \frac{T_{c_i}}{2}\right) T_{c_i}\right) \sin(\theta_{e_i}) + n_{Q_i} \tag{4.2}$$

where τ_{e_i} is the code-delay error at the ith interval; f_{e_i} is the Doppler shift estimation error at the beginning of the ith interval; α_e is the Doppler rate estimation error; θ_{e_i} is the average phase error over the ith interval; and T_{c_i} is the C/A code length modified by the Doppler effect at the ith interval. This chapter and the next use seconds as the unit of time and hertz as the unit of the Doppler shift.

The developed algorithms are formulated to work with signals under any dynamic condition. A change in the Doppler shift causes a change in the code length, so the code and the data bit lengths are not the same in each interval. The lengths are defined by T_{c_i} and T_{d_m}, respectively. The index i refers to the ith C/A code length, and the index m refers to the mth data bit length.

If a received signal has low dynamics, then both the received code and the data bit are assumed to have constant lengths. In this case, these lengths are calculated at the start of each algorithm and then used throughout its processing.

These values are defined as T_{c0} and T_{d_0} for the received code and data bit lengths, respectively. They are calculated from

$$T_{c0} = T_c \frac{f_{L1}}{f_{L1} + \hat{f}_{d_0}} \tag{4.3}$$

$$T_{d_0} = T_d \frac{f_{L1}}{f_{L1} + \hat{f}_{d_0}} \tag{4.4}$$

where $\hat{f}_{d_0}$ is the estimated Doppler shift at the start of an algorithm; and T_c and T_d are the actual code and data bit lengths, which are equal to 0.001 sec. and 0.02 sec. for the GPS C/A signal, respectively. So,

$$T_{c_i} = T_{c0} \quad i = 1, 2, \ldots \tag{4.5}$$

$$T_{d_m} = T_{d_0} \quad m = 1, 2, \ldots \tag{4.6}$$

In the case of high dynamics, it is not necessary to recalculate T_{c_i} and T_{d_m} in every step. They are calculated after a certain number of update steps; this number of steps is based on the Doppler rate value. This is because the algorithms developed in this chapter do not require very accurate code and data bit lengths. The accurate lengths are required in the code- and carrier-tracking modules, introduced in Chapter 5, Sections 5.3 and 5.4, which operate over a long period of time.

4.3 Acquisition, Fine Acquisition, and Bit Synchronization Modules Relationship

Figure 4.1 shows the relationship between the acquisition, fine acquisition, and bit synchronization modules. The output of the acquisition algorithms developed in Chapter 3 consists of approximate code delay, Doppler shift, Doppler rate, and a few possible bit edge positions; this output is passed to the signal generator. The signal generator generates the Doppler-compensated local replica code; taking into account the estimates of the carrier parameters. It also aligns the local replica code with the received one using the code-delay estimate, and it passes 1-ms correlated signals to the fine acquisition module. The acquisition module passes only the approximate bit edge positions to the fine acquisition module.

The fine acquisition module calculates estimates for the carrier parameters errors; the output is used to update the signal generator estimates. The signal generator passes updated 1-ms signals to the bit synchronization module, which uses the VA/EKF. The VA/EKF assumes that the initial estimates of the carrier

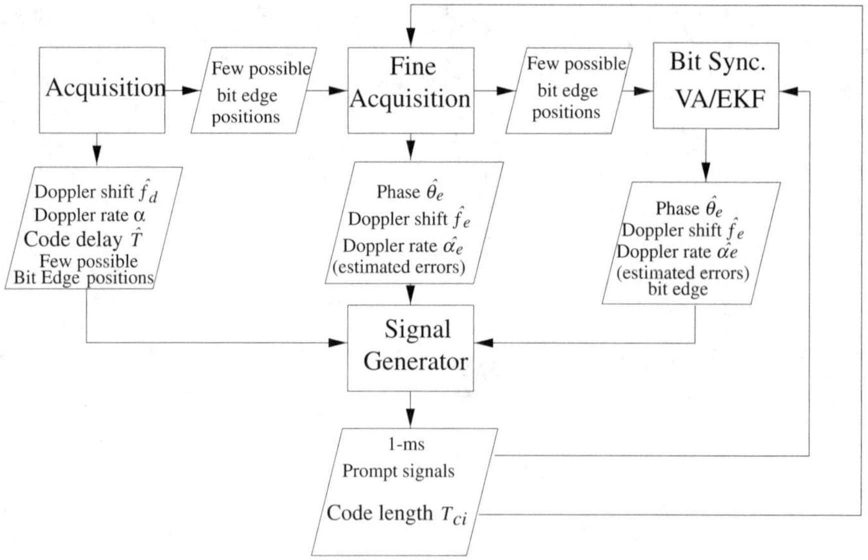

Figure 4.1 The relationships between acquisition, fine acquisition, bit synchronization, and the signal generator.

parameters errors are zeros, then estimates the bit edge position and data bit sequence and tracks the carrier parameters errors. During the bit synchronization operation, the signal generator does not update its estimates of the carrier parameters. After the bit edge position is estimated, the VA/EKF passes the estimated bit edge position and the updated estimates of the carrier parameters errors to the signal generator, which uses them to update its estimates. Then, the signal generator generates the prompt, early, and late signals with integration of one data bit interval starting at the estimated bit edge position. These signals are passed to the code- and carrier-tracking modules, which are presented in Chapter 5. The carrier-tracking module performs tracking on the actual phase, Doppler shift, and Doppler rate, not on the errors in these parameters like the tracking performed during the bit synchronization.

4.4 Bit Synchronization and Data Bit Detection Algorithm

The problem of estimating the bit edge position and the data bit values is formulated to allow for the application of the VA. Assume first that there are no carrier- or code-tracking errors. Thus, only the in-phase correlated signals are considered. Let $N+1$ define the number of data bit intervals that are used in the algorithm. Each data bit interval has 20 T_{c_i}-sec. correlated signals. There are at most $N+2$

data bits in the $N+1$ data intervals, T_{d_m}; there are at least N complete data bits in the same intervals. The model of the T_{c_i}-sec. in-phase correlated signals follows from (4.1), as

$$I_i = A\, d_i + n_{I_i} \tag{4.7}$$

Since d_i has the same value for each 20 consecutive signals belonging to the same data bit, define the sequence of the received data bits as $\{d_\nu\}_{\nu=1}^{N+2}$ and e as the number of the T_{c_i}-sec. signals missed from the first received data bit. The observations are formed such that each represents the average of the summation of 20 consecutive T_{c_i}-sec. signals; N consecutive observations are formed. Since there are 20 possible bit edge positions, then 20 different sets of observations are formed. For each set, the summation starts at an index ϕ equal to a possible bit edge position, $\phi = 1, \ldots, 20$. Each observation is defined by $s_{m,\phi}$, where m is the observation number, and ϕ is the set number. This is represented as

$$s_{m,\phi} = \frac{1}{\sqrt{20}} \sum_{i=20\,(m-1)+\phi}^{20\,m+\phi-1} I_i \tag{4.8}$$

where

$$\sum_{i=20\,(m-1)+\phi}^{20\,m+\phi-1} I_i = \left\{ \sum_{\nu=20\,(m-1)+\phi}^{20\,m-e} A\, d_\nu + \sum_{\nu=20\,m-e+1}^{20\,m+\phi-1} A\, d_{\nu+1} \right\} + \sum_{i=20\,(m-1)+\phi}^{20\,m+\phi-1} n_{I_i} \tag{4.9}$$

The division by $\sqrt{20}$ is done to keep the noise variance equal to 1. The observations are expressed as

$$s_{m,\phi} = A_{m,\phi}\, d_{m,\phi} + n_{m,\phi} \tag{4.10}$$

where $A_{m,\phi}$ is the average signal level over the mth observation interval, and $n_{m,\phi}$ is the average noise. If ϕ is not the correct bit edge position, then $d_{m,\phi}$ and $A_{m,\phi}$ are found from averaging over two data bits. If the two data bits have different signs, then $A_{m,\phi}$ is found from

$$A_{m,\phi} = \frac{1}{\sqrt{20}} \left[20 - 2(\mathit{EdgeNum} - 1)\right] A \tag{4.11}$$

where *EdgeNum* is the index of the location of ϕ within a data bit such that

$$\mathit{EdgeNum} = \begin{cases} e + \phi & \text{if } \phi \le 20 - e \\ e + \phi - 20 & \text{if } \phi > 20 - e \end{cases}$$

Also, if the two data bits have different signs, then the sign of $d_{m,\phi}$ is the same as the sign of the data bit that has larger number of T_{c_i}-sec. samples within the mth interval. If ϕ is not the correct bit edge position and the two averaged data bits have the same sign, or if ϕ is the correct bit edge position, then $A_{m,\phi}$ is found from

$$A_{m,\phi} = \sqrt{20}\, A$$

If ϕ is the correct bit edge position, then

$$d_{m,\phi} = \begin{cases} d_\nu & \text{if } e = 0 \\ d_{\nu+1} & \text{if } e \neq 0 \end{cases}$$

Thus, $s_{m,\phi}$ will be maximum if ϕ is the correct bit edge position, or if the average is done over two data bits that have the same sign. Otherwise, $s_{m,\phi}$ will be degraded. However, the problem of estimating the data bit values and the bit edge position for weak signals is that the noise, $n_{m,\phi}$, degrades $s_{m,\phi}$. So, a maximum likelihood (ML) approach is applied for the estimation. Assume that the samples I_i are normally distributed, and so the sequence $\{s_{m,\phi}\}_{m=1}^{N}$ is too; n_i is AWGN with zero mean and unit variance, and so $n_{m,\phi}$ is too; $\{d_m\}_{m=1}^{N}$ is an independent, equally likely data sequence, with values of $+1$ or -1. Given the sequence $\{s_{m,\phi}\}_{m=1}^{N}$, for $\phi = 1, \ldots, 20$, the ML estimation of the bit edge position and the data bit values can be found by solving the maximization

$$\max_{\{d_m\}_1^N,\, \phi} f(\mathbf{S}|\{d_m\}_1^N,\, \phi) \tag{4.12}$$

where $f(\mathbf{S}|\{d_m\}_1^N,\, \phi)$ is the likelihood function of $\mathbf{S}$; it is expressed in the form

$$f(\mathbf{S}|\{d_m\}_1^N,\, \phi) = \frac{1}{(2\pi\,\sigma_n^2)^{\frac{N}{2}}} \exp\left(-\frac{1}{2\sigma_n^2}\sum_{m=1}^{N} |s_{m,\phi} - \bar{A}\, d_m|^2\right) \tag{4.13}$$

where $\bar{A}$ is the expected signal level, given that ϕ is the correct bit edge position. The likelihood function in (4.13) is replaced by its log likelihood, so

$$\ln f(\mathbf{S}|\{d_m\}_1^N,\, \phi) = -\, B \sum_{m=1}^{N} |s_{m,\phi} - \bar{A}\, d_m|^2 \tag{4.14}$$

where B is a constant. Thus, the maximization in (4.12) is equivalent to

$$\max_{\{d_m\}_1^N,\, \phi} \; - \sum_{m=1}^{N} |s_{m,\phi} - \bar{A}\, d_m|^2 \tag{4.15}$$

This is equivalent to the minimization

$$\min_{\{d_m\}_1^N,\,\phi} \sum_{m=1}^{N} |s_{m,\phi} - \bar{A}\, d_m|^2 \tag{4.16}$$

This can be expressed in a recursive form by defining $\Gamma_{N,\phi}$ as

$$\Gamma_{N,\phi} = \sum_{m=1}^{N} |s_{m,\phi} - \bar{A}\, d_m|^2 \tag{4.17}$$

This satisfies the recursion

$$\Gamma_{m,\phi} = \Gamma_{m-1,\phi} + p_{m,\phi}, \quad m = 2, 3, \ldots, N \tag{4.18}$$

where

$$p_{m,\phi} = |s_{m,\phi} - \bar{A}\, d_m|^2 \tag{4.19}$$

$$\Gamma_{1,\phi} = |s_{1,\phi} - \bar{A}\, d_1|^2 \tag{4.20}$$

where $p_{m,\phi}$ is taken as a weight function for the mth observation. The optimal estimation for the bit edge position and the data bit values, at a step m, will be found to be those values that generate the minimum $\Gamma_{m,\phi}$. Thus, it is concluded that the optimal solution can be found from dynamic programming; thus, the VA is applied to solve the minimization in (4.16).

4.4.1 Minimization Solution

The minimization in (4.16) and the recursion in (4.18) are solved as follows. Each ϕ is represented by a separate state diagram. Each state diagram has two states representing the possible data bit values ± 1, and each is fully connected, so there are four transitions, as shown in Figure 4.2. Figure 4.3 is derived from Figure 4.2

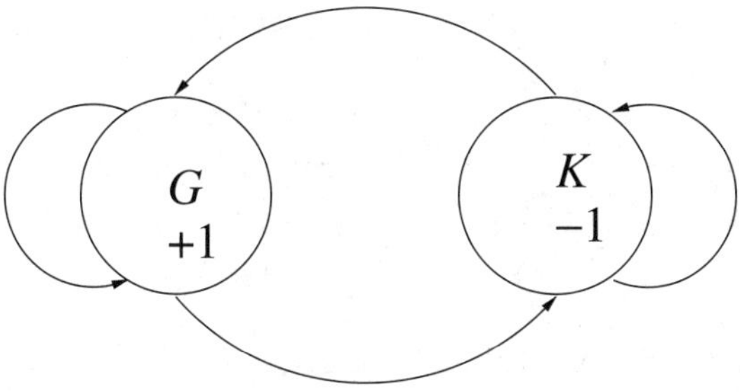

Figure 4.2 State diagram G and K represent the data bit values (± 1).

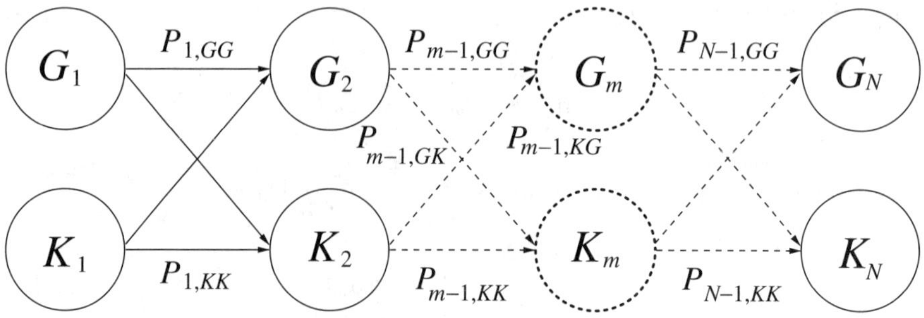

Figure 4.3 Trellis graph derived from the state diagram by following all possible transitions through N steps.

by following all of the possible transitions through N steps. Each arrow represents a possible transition from step m to step $m+1$; (4.19) is used to calculate the path weight for each transition.

Let $p_{m,\phi,hl}$ define the path weight connecting a state h at the mth step to a state l at the $(m+1)$st step; in figure 4.3, $h = G, K$ and $l = G, K$. Let, $\Gamma_{m,\phi,h}$ defines the cumulative weight of a state h at the mth step. $\Gamma_{m,\phi,hl}$ defines the cumulative weight resulting from moving from a state h at the mth step to a state l at the $(m+1)$st step. At each step, $m+1$, the following are performed:

(1) For each state h at the mth step, $\Gamma_{m,\phi,hl} = \Gamma_{m,\phi,h} + p_{m,\phi,hl}$ is calculated.

(2) For each state l at the $(m+1)$st step, the minimum $\Gamma_{m,\phi,hl}$, $h = G, K$, is found.

(3) $\Gamma_{m+1,\phi,l}$ is set to be equal to this minimum value, and the data bit estimation for a state l is updated such that the data bit sequence estimation from $d_{1,\phi}$ to $d_{m,\phi}$ is equal to the sequence estimation associated with the minimum $\Gamma_{m,\phi,hl}$. The estimation of the data bit $d_{m+1,\phi}$ is chosen according to the state l. After each step, each state will keep track of only one sequence, which is the one that generates the minimum cumulative weight up to this step.

After each step, each possible ϕ will have two cumulative weights resulting from the states K and G. Let $\Gamma_{m,\phi} = \min\{\Gamma_{m,\phi,G}, \Gamma_{m,\phi,K}\}$. After N steps, each possible ϕ will have a $\Gamma_{N,\phi}$ that represents the minimum cumulative path weight associated with that edge index. The minimum of $\Gamma_{N,\phi}$, for $\phi = 1, \ldots, 20$, will represent the estimation of the bit edge position, and its associated path will represent the estimation of the data bit sequence.

4.5 Extended States VA for Fine Acquisition

The acquisition process provides estimates for the Doppler shift and Doppler rate. The accuracy of the estimation depends on the PIT used in that process. Generally, such accuracy will need to be refined before activating a carrier-tracking module. A carrier-tracking module that uses an EKF is developed in Chapter 5. In order for the EKF to converge, it has to have an initial Doppler frequency error less than about 12 Hz, a small initial Doppler rate error, and a phase estimate within 45° accuracy. A modified version of the VA, the extended states VA (ES-VA), is used to provide the fine estimation.

Assume first that the Doppler rate error, α_e, is negligible; thus, $\hat{\alpha} \approx \alpha$. The in-phase and the quad-phase T_{c_i}-sec. signals in (4.1) and (4.2) will have the following form:

$$I_i = A\, d_i\, \operatorname{sinc}(f_{e_i}\, T_{c_i})\, \cos(\theta_{e_i}) + n_{I_i} \tag{4.21}$$

$$Q_i = A\, d_i\, \operatorname{sinc}(f_{e_i}\, T_{c_i})\, \sin(\theta_{e_i}) + n_{Q_i} \tag{4.22}$$

where

$$f_{e_i} = f_{e_{i-1}} + \alpha_e\, T_{c_i} + W_{f_d,i} \tag{4.23}$$

$$\theta_{e_i} = \theta_{e_{i-1}} + 2\pi\, f_{e_i}\, T_{c_i} + W_{\theta,i} \tag{4.24}$$

where $W_{f_d,i}$ and $W_{\theta,i}$ are the clock frequency and phase disturbances, respectively. If the clock disturbances are small as compared to the frequency and phase errors, and if $\alpha_e \approx 0$, then (4.23) and (4.24) are approximated by

$$f_{e_i} \approx f_{e_0} \tag{4.25}$$

$$\theta_{e_i} \approx \theta_{e_0} + 2\pi\, f_{e_0} \left(\sum_{n=0}^{i} T_{c_n} \right) \tag{4.26}$$

or, in the case of low dynamics,

$$\theta_{e_i} \approx \theta_{e_0} + 2\pi\, f_{e_0}\, (i\, T_{c_0}) \tag{4.27}$$

where f_{e_0} and θ_{e_0} are the frequency and phase errors at $i = 0$. From (4.21) and (4.22), define y_i as

$$y_i = A\, d_i\, \operatorname{sinc}(f_{e_i}\, T_{c_i})\, e^{j\theta_{e_i}} + n_i \tag{4.28}$$

Define f_t as a possible frequency error; (4.8) is modified to

$$s_{m,\phi,f_t} = \frac{1}{\sqrt{20}} \sum_{i=20(m-1)+\phi}^{20m+\phi-1} y_i \, e^{-j2\pi f_t \left(\sum_{n=\phi}^{i} T_{cn}\right)} \tag{4.29}$$

$$s_{m,\phi,f_t} = \frac{1}{\sqrt{20}} \sum_{i=20(m-1)+\phi}^{20m+\phi-1} \Bigg\{ A\, d_i \, \mathrm{sinc}(f_{e_0} T_{c_i}) \, e^{j\left[\theta_{e_0} + 2\pi (f_{e_0} - f_t)\left(\sum_{n=\phi}^{i} T_{cn}\right)\right]} + n_i \, e^{-j2\pi f_t \left(\sum_{n=\phi}^{i} T_{cn}\right)} \Bigg\} \tag{4.30}$$

or, in the case of low dynamic,

$$s_{m,\phi,f_t} = \frac{1}{\sqrt{20}} \sum_{i=20(m-1)+\phi}^{20m+\phi-1} \Big\{ A\, d_i \, \mathrm{sinc}(f_{e_0} T_{c0}) \, e^{j[\theta_{e_0} + 2\pi (f_{e_0} - f_t)(i\, T_{c0})]} + n_i \, e^{-j2\pi f_t (i\, T_{c0})} \Big\} \tag{4.31}$$

Equations (4.16) and (4.17) are modified to

$$\min_{\{d_m\}_1^N,\, \phi,\, f_t} \sum_{m=1}^{N} \left| s_{m,\phi,f_t} - \bar{A}\, d_m \right|^2 \tag{4.32}$$

$$\Gamma_{N,\phi,ft} = \sum_{m=1}^{N} \left| s_{m,\phi,f_t} - \bar{A}\, d_m \right|^2 \tag{4.33}$$

Thus, (4.18) and (4.20) are applied to find an estimate for f_{e_0}. The VA states are extended to account for N_{freq} possible frequency errors, the tested frequencies f_t. Thus, the number of VA states will be increased by a factor of N_{freq}. The received T_{c_i}-sec. signals are counterrotated by each of those frequencies, as in (4.29), to generate N_{freq} streams of signals from the original received one. Each stream is used to generate the observations s_{m,ϕ,f_t}. The VA then proceeds as described in Section 4.4; at the end of N steps, the path that has the minimum cumulative sum $\Gamma_{N,\phi,ft}$ will correspond to the best estimates of the frequency error, the bit edge position, and the data bit sequence.

Using a large number of N_{freq} will increase processing and memory requirements. Therefore, the frequency is estimated iteratively, and a small number of N_{freq} is used in each iteration. The algorithm starts with a search range determined from the maximum and the minimum expected frequency errors, with large separation between the tested frequencies. After each iteration, a rough estimate of the frequency error is generated, and then the search range and the frequency separation are reduced. This is done as follows. The maximum and the minimum expected frequency errors are defined by f_{max} and f_{min}, respectively. The separation between the tested frequencies in the jth iteration is defined by f_{step_j}. The size of the frequency error range in the jth iteration is defined by f_{range_j}. The start and the end frequency values of f_{range_j} are defined by f_{start_j} and f_{end_j}, respectively. The first start and end frequency values, f_{start_1} and f_{end_1}, are set equal to f_{min} and f_{max}, respectively. The size of the search range of the jth iteration is

$$f_{range_j} = f_{end_j} - f_{start_j} \tag{4.34}$$

The frequency separation is calculated by

$$f_{step_j} = \left\lceil \frac{f_{range_j}}{N_{freq} - 1} \right\rceil \tag{4.35}$$

The ES-VA then proceeds to find the test frequency value, f_{est_j}, that generates the minimum cumulative sum. The search range is then reduced by a factor $R_{reduction}$. So, $f_{range_{j+1}}$ is calculated from f_{range_j} as

$$f_{range_{j+1}} = f_{range_j} \cdot R_{reduction} \tag{4.36}$$

The frequency separation, $f_{step_{j+1}}$, is then calculated as in (4.35). f_{est_j} is used as the middle frequency of the new search range, $f_{range_{j+1}}$. The start and the end search frequency values for the $(j+1)$th iteration are

$$f_{start_{j+1}} = \max\left(f_{est_j} - \frac{f_{range_{j+1}}}{2},\ f_{start_j} \right) \tag{4.37}$$

$$f_{end_{j+1}} = \min\left(f_{est_j} + \frac{f_{range_{j+1}}}{2},\ f_{end_j} \right) \tag{4.38}$$

The algorithm continues until $f_{step_{j+1}}$ is less than a minimum separation determined by a predefined value, $f_{resolution}$. If J defines the number of iterations, then f_{est_J} is taken as the frequency error estimate. Since it is assumed that $f_{e_i} \approx f_{e_0}$, then

$$\hat{f}_{e_i} = f_{est_J} \tag{4.39}$$

It should be noted that f_{est_J} represents the frequency error estimate at the beginning of the interval that is used in the last iteration of the ES-VA. $\hat{f}_{e_i}$ is then used to update the Doppler shift estimate.

Since the ES-VA is not intended to find the bit edge position, it is not necessary to construct an observation set from each possible bit edge position. Instead, the ES-VA can search number of edges, N_{edge}, equal to

$$N_{edge} = \left\lceil \frac{N_{sBit}}{S_{separation}} \right\rceil \tag{4.40}$$

where N_{sBit} is the number of signals in the observation interval (i.e., 20 T_{c_i}-sec. signals). $S_{separation}$ is the chosen separation between the searched edges. The algorithm can use a different number of bit intervals, N, depending on the C/N_0 and the required accuracy for the initial frequency. Also, a small number of bit intervals can be used when the frequency separation is large; then, as the separation decreases, the number of bit intervals can be increased.

A similar method is used to refine the estimated phase, $\hat{\theta}$, which is also needed to initialize the tracking module. However, the phase error estimate needs very few numbers of iterations. The method uses the previous estimated frequency f_{est_J}, along with tested phase values θ_t. Only the rotated in-phase part of the signal (i.e., $\Re\{y\, e^{-j\,(\theta_t + 2\pi\, f_{est_J}\, t)}\}$) is used to form the measurement, $s_{m,\phi,\,\theta_t}$. Since the method proceeds without any knowledge of the data bit values, then the estimated phase is expected to be close to the correct phase or the phases that are separated by π from the correct one. If both the rotated in-phase and quad-phase signals are used, then the estimated phase value will be close to the correct phase or the phases that are separated by $\pi/2$, $-\pi/2$, or π from the correct one. The first search range is $(-\pi, \pi)$. After the first iteration, the method proceeds by forming a narrow search range with the last estimated phase in the middle of the range. This method is sensitive to the errors in the frequency estimation, f_{est_J}, so it is not very robust.

4.6 High Dynamic Fine Acquisition

In Section 4.5, the ES-VA is used to refine the estimated f_d and θ, but α_e is assumed to be small, so no additional steps are needed. To obtain refined estimates for f_d and α, a new version of the ES-VA is used without the need to consider all of the possible combinations of f_d and α errors.

The estimation is done in two steps. In the first step, the frequency error is assumed to be zero, and then the ES-VA is applied to obtain an estimate for the Doppler rate error, α_{f_e}, where α_{f_e} is a function in the true frequency error, f_e, and the true Doppler rate error, α_e. In the second step, the relationship between

α_{f_e}, f_e, and α_e is used to obtain estimates for f_e and α_e. The estimation is done using an approach similar to the one described in Section 4.5.

In the first step, the estimation is done iteratively using a small number of additional states and starting with a large separation between the tested Doppler rates. As the algorithm progresses, the separation is reduced. α_{f_e} is concluded when the separation reaches a predefined value of $\alpha_{resolution}$ Hz/s. Define T_α as the amount of time used in the last iteration of the ES-VA to obtain α_{f_e} (i.e., $T_\alpha = \sum_{m=1}^{N} T_{d_m}$). Define f_{e_0} as the frequency error at the beginning of the T_α interval. The true f_{e_0} and α_e are related to α_{f_e} by

$$\alpha_{f_e} = C\frac{f_{e_0}}{T_\alpha} + \alpha_e + E_\alpha \tag{4.41}$$

where $C = 3/2$ is a constant, and E_α is an estimation error.

In the second step of the estimation, three methods are introduced to obtain the true estimates of α_e and f_{e_0}. The first one, frequency estimate with zero rate (FEZR), applies the ES-VA under the assumption that α_e is zero. In this case, the ES-VA generates a frequency estimate, f_{est_J}, that maintains the following relationship:

$$f_{est_J} = f_{e_0} + \alpha_e \frac{T_f}{2} + E_f \tag{4.42}$$

where T_f is the amount of time used to obtain f_{est_J} in the last iteration of the ES-VA, f_{e_0} is the true frequency error at the beginning of the T_f interval; and E_f is an estimation error.

From (4.41) and (4.42), and assuming that the start of the T_α interval is the same as the start of the T_f interval, the true frequency error and the Doppler rate error are calculated from

$$f_{e_0} = \left[\frac{2T_\alpha}{2T_\alpha - CT_f}\right]\left[f_{est_J} - \frac{T_f}{2}\alpha_{f_e} + \frac{T_f}{2}E_\alpha - E_f\right] \tag{4.43}$$

$$\alpha_e = \alpha_{f_e} - C\frac{f_{e_0}}{T_\alpha} - E_\alpha \tag{4.44}$$

However, the problem is that both E_α and E_f are unknown. Therefore, an additional step is needed to find the required estimates. In this case, different values for f_{e_0} and α_e are calculated using different possibilities for the errors. Then, the ES-VA is run for one step using those values. For example, if the errors are taken as -0.5, 0, and 0.5 for both E_f and E_α, then the ES-VA will run with nine different combinations of f_{e0} and α. However, if the number of possible errors is increased, the number of the states in the ES-VA will also increase. The number of the states is equal to $N_{E_f}\ N_{E_\alpha}$. N_{E_f} is the number of possible errors in the Doppler shift estimation. N_{E_α} is the number of possible errors in the Doppler rate estimation.

The second method to obtain the true estimates, frequency estimate with estimated rate (FEER), uses α_{f_e} directly in the ES-VA when estimating the frequency error. In this case, (4.42) is modified as

$$f_{est_J} = f_{e0} + (\alpha_e - \alpha_{f_e})\,\frac{T_f}{2} + E_f \tag{4.45}$$

From (4.41) and (4.45), the frequency error is

$$f_{e0} = \left[\frac{2\,T_\alpha}{2\,T_\alpha - C\,T_f}\right]\left[f_{est_J} + \frac{T_f}{2}\,E_\alpha - E_f\right] \tag{4.46}$$

A similar problem exists in FEER as in FEZR due to the unknowns E_α and E_f. Therefore, an additional step is needed to obtain the final estimate, as in the first method.

The third method, frequency and rate estimate (FRE), uses (4.41) to calculate a tested rate, α_t, for each tested frequency, f_t, in the ES-VA as

$$\alpha_t = \alpha_{f_e} - C\,\frac{f_t}{T_\alpha} - E_\alpha \tag{4.47}$$

This method will generate estimates for both f_{e0} and α_e, so no additional calculation is needed. However, to deal with the unknown E_α, the following steps are taken. At each iteration of the ES-VA, for each tested frequency, a number N_{E_α} of possible Doppler rate errors is considered. The N_{E_α} values are calculated from (4.47), with different values for E_α. The number of the states is equal to $N_{freq}\ N_{E_\alpha}$. After each iteration of the ES-VA, a frequency estimation is found and used as the middle range in the following iteration; the α_t associated with the estimated frequency will not affect the choice of the f_t values. After the ES-VA is concluded, the estimated f_{e0} and its associated α_e will be taken as these values' estimates at the last step of the ES-VA.

Any of the three methods (FEZR, FEER, and FRE) can be used to estimate the frequency and the Doppler rate errors. After that, the ES-VA is used for phase error estimation, where both the frequency and the Doppler rate estimated errors are used in this process. The three estimated values are used to update the phase, Doppler shift, and Doppler rate estimates.

4.7 Bit Synchronization and Navigation Message Detection in the Presence of Carrier-Tracking Errors

Section 4.4 develops a bit synchronization and data bit detection algorithm, which assumes that there are no phase or frequency errors. This section, modifies this algorithm to deal with the existence of those errors. Thus, estimates for the bit edge position and the data bit sequence are obtained under weak-signal conditions.

An EKF is used in the VA paths to track the phase, frequency, and Doppler rate errors. Each observation set, of the 20 sets, operates a separate EKF, which is updated before each observation processing. The result is used to reduce the estimation errors by counterrotating the signals of the following observation according to the current estimated phase, frequency, and Doppler rate errors. The concept of estimating the phase error and counterrotating the received signal with that error is common in communication theory [3], where the phase tracking is performed with a first- or second-order loop. Generally, second-order loops are used for the phase tracking of the VA survivor paths [4–6]. The EKF has some advantages over the tracking loops since it includes a model for line-of-sight dynamics, and it updates its estimates taking into account the measurement noise and the uncertainty in its current estimates, in contrast to the tracking loops. Also, the ability of a loop to track the line-of-sight dynamics depends on its order. Higher-order loops are complicated and need to be carefully designed, whereas the same EKF structure can easily be extended to more complex dynamic models.

4.7.1 VA with an EKF

An EKF is used to track the phase, Doppler shift, and Doppler rate. The EKF output is used to provide updated estimates to the local signal generator module after each EKF update step. Thus, the correlation loss due to the errors in these parameters, which is expressed by the sinc function in (4.1) and (4.2), will be updated continuously. The output phase, Doppler shift, and Doppler rate errors in the correlated signal at the ith interval are equal to

$$\theta_{e_i} = \theta_i - \hat{\theta}_i \tag{4.48}$$

$$f_{e_i} = f_{d_i} - \hat{f}_{d_i} \tag{4.49}$$

$$\alpha_{e_i} = \alpha - \hat{\alpha}_i \tag{4.50}$$

The developed bit synchronization algorithm uses a separate EKF for each possible bit edge position. After each EKF update step, there will be 20 different estimated values for each of the carrier parameters. It is impractical to have 20 different local signal generators, one for each possible bit edge, because this will increase the processing requirements. The ES-VA provides fine estimates of the carrier parameters; thus, its output is used to provide updated estimates to the local signal generator. This will reduce the correlation loss. The signal generator estimates of the carrier parameters are not updated during the bit synchronization process. After the bit edge position is identified, the EKFs at the incorrect edge positions will be disabled. The EKF output at the correct bit edge position will be used to provide updated estimates to the local signal generator.

If the EKF output is not used to provide updated estimates to the local signal generator module, then the output Doppler rate, Doppler shift, and phase errors on the correlated signals in the ith interval will be equal to

$$\alpha_{e_i} = \alpha_{e_0} \tag{4.51}$$

$$f_{e_i} = f_{e_{i-1}} + \alpha_{e_0} T_{c_i} + W_{f_d,i} \tag{4.52}$$

and

$$\theta_{e_i} = \theta_{e_{i-1}} + 2\pi f_{e_i} T_{c_i} + W_{\theta,i} \tag{4.53}$$

The EKF is used to update the local signal generator in Chapter 5, Sections 5.3 and 5.4; it is implemented to track the actual phase, Doppler shift, and Doppler rate. In this chapter, the EKF output is not used to update the signal generator; it is implemented to track the errors in the correlator output signals. It should be noted, however, that the EKF can be used to track the actual values of the carrier parameters, even if its output is not used to update the local signal generator. The problem in this case is that the values in (4.51) to (4.53) will need to be calculated after each step, which will add unnecessary processing. It is very easy to switch the EKF tracking from one mode to the other, which is done once the bit edge is identified.

4.7.2 Extended Kalman Filter Model

Let θ, ω, and α_ω define, respectively, the phase, the radian frequency, and the radian frequency rate that are tracked by an EKF. The state vector, x, is defined as

$$x = \begin{bmatrix} \theta \\ \omega \\ \alpha_\omega \end{bmatrix}$$

where $\omega = 2\pi f$, $\alpha_\omega = 2\pi\alpha$, and

$$\dot{x} = F\,x + W$$

This is found as

$$\begin{bmatrix} \dot{\theta} \\ \dot{\omega} \\ \dot{\alpha}_\omega \end{bmatrix} = \begin{bmatrix} 0 & 1 & 0 \\ 0 & 0 & 1 \\ 0 & 0 & 0 \end{bmatrix} \begin{bmatrix} \theta \\ \omega \\ \alpha_\omega \end{bmatrix} + \begin{bmatrix} W_\theta \\ W_\omega \\ W_{\alpha_\omega} \end{bmatrix}$$

where W_θ, W_ω, and W_{α_ω} define the noise disturbances of the phase, the frequency, and the frequency rate, respectively. The transition matrix, Φ, is derived from the Taylor series as

$$\Phi = I + F\,t + \frac{1}{2}\,F^2\,t^2$$

where,

$$F = \begin{bmatrix} 0 & 1 & 0 \\ 0 & 0 & 1 \\ 0 & 0 & 0 \end{bmatrix}$$

t is a continuous time and is replaced by T_{d_m}, which is the measurement (observation) interval. This leads to the following dynamics model:

$$x_{m+1} = \Phi x_m + W_{x,m+1} \tag{4.54}$$

$$\begin{bmatrix} \theta_{m+1} \\ \omega_{d_{m+1}} \\ \alpha_{\omega_{m+1}} \end{bmatrix} = \begin{bmatrix} 1 & T_{d_m} & \frac{T_{d_m}^2}{2} \\ 0 & 1 & T_{d_m} \\ 0 & 0 & 1 \end{bmatrix} \begin{bmatrix} \theta_m \\ \omega_{d_m} \\ \alpha_{\omega_m} \end{bmatrix} + \begin{bmatrix} W_{\theta,m+1} \\ W_{\omega,m+1} \\ W_{\alpha_\omega,m+1} \end{bmatrix} \tag{4.55}$$

This dynamics model assumes constant acceleration, and it is linear, so the state transition matrix is exact. From (2.40) and (2.41), the phase and frequency disturbances are modeled as normal random walks, each with zero mean. The disturbances are assumed to be independent. The variances are defined as

$$E[W_\theta\, W_\theta^T] = Q_\theta$$

$$E[W_\omega\, W_\omega^T] = Q_\omega$$

$$E[W_{\alpha_\omega}\, W_{\alpha_\omega}^T] = Q_{\alpha_\omega}$$

This leads to the following process noise matrix:

$$Q_n = \begin{bmatrix} Q_\theta & 0 & 0 \\ 0 & Q_\omega & 0 \\ 0 & 0 & Q_{\alpha_\omega} \end{bmatrix}$$

Since the algorithm processes discrete observations, the discrete process noise is found using the transition matrix [7] as

$$Q_{n_m} = \int_0^{T_{d_m}} \Phi(t)\, Q_n\, \Phi^T(t)\, dt$$

So,

$$Q_{n_m} = \begin{bmatrix} Q_\theta\, T_{d_m} + Q_\omega \frac{T_{dm}^3}{3} + Q_{\alpha_\omega} \frac{T_{dm}^5}{20} & Q_\omega \frac{T_{dm}^2}{2} + Q_{\alpha_\omega} \frac{T_{dm}^4}{8} & Q_{\alpha_\omega} \frac{T_{dm}^3}{6} \\ Q_\omega \frac{T_{dm}^2}{2} + Q_{\alpha_\omega} \frac{T_{dm}^4}{8} & Q_\omega\, T_{d_m} + Q_{\alpha_\omega} \frac{T_{dm}^3}{3} & Q_{\alpha_\omega} \frac{T_{dm}^2}{2} \\ Q_{\alpha_\omega} \frac{T_{dm}^3}{6} & Q_{\alpha_\omega} \frac{T_{dm}^2}{2} & Q_{\alpha_\omega}\, T_{d_m} \end{bmatrix}$$

The measurement noise is defined as

$$R_n = \sigma_n^2$$

4.7.3 EKF Estimation

The phase estimation using an EKF is formulated as follows. The signal model is taken as in (4.28). The summation over 20 of the T_{c_i}-sec. signals, defined in (4.10), is modified as

$$s_{m,\phi} = A_{m,\phi}\, d_{m,\phi}\, e^{j\,\theta_{e_{m,\phi}}} + n_{m,\phi} \tag{4.56}$$

where $\theta_{e_m,\phi}$ is the average phase error over the mth observation interval for a possible bit edge ϕ. $A_{m,\phi}$ is the average of the signal amplitude multiplied by the sinc function over the observation interval.

The phase estimation over one observation interval is expressed with a log-likelihood function similar to (4.14), as

$$\Lambda(\mathbf{Y}|\,\theta_{e_m,\phi}) = \ln f(\mathbf{Y}|\,\theta_{e_m,\phi}) = -\,C \sum_{i=20\,(m-1)+\phi}^{20\,m+\phi-1} \left| y_i - \bar{A}\, d_i\, e^{j\,\theta_{e_i}} \right|^2 \tag{4.57}$$

where C is a constant. This is expanded, and the irrelevant constants are discarded to give a log likelihood, Λ, as

$$\Lambda\left(\mathbf{Y}|\,\theta_{e_m,\phi}\right) = \Re\left\{ d_{m,\phi} \sum_{i=20\,(m-1)+\phi}^{20\,m+\phi-1} y_i\, e^{-\,j\,\theta_{e_i}} \right\} \tag{4.58}$$

where $\Re$ refers to the real part of the function inside brackets in (4.58). The phase estimation is found by computing the zero of the derivative of the log-likelihood function [3], and replacing $d_{m,\phi}$ and θ_{e_i} with the estimates $\hat{d}_{m,\phi}$ and $\hat{\theta}_{e_i}$, respectively. So,

$$\frac{d}{d\,\theta_{e_m,\phi}}\,\Lambda\left(\mathbf{Y}|\,\theta_{e_m,\phi}\right) = \hat{d}_{m,\phi} \sum_{i=20\,(m-1)+\phi}^{20\,m+\phi-1} \Im\left\{ y_i\, e^{-\,j\,\hat{\theta}_{e_i}} \right\} \tag{4.59}$$

where $\Im$ refers to the imaginary part of the function inside bracket in (4.59). From (4.28) and (4.56), (4.59) is approximated by

$$\frac{d}{d\,\theta_{e_m,\phi}}\,\Lambda\left(\mathbf{Y}|\,\theta_{e_m,\phi}\right) \approx \hat{d}_{m,\phi}\; d_{m,\phi}\; A_{m,\phi}\; \sin\left(\theta_{e_m,\phi} - \hat{\theta}_{e_m,\phi}\right) \tag{4.60}$$

Based on this derivation, the residual, Res_m, and the linearization function, H_m, of the EKF are derived to give a value proportional to the phase error.

The VA uses a separate EKF for each bit edge, ϕ. The EKF estimates the average phase, frequency, and frequency rate over an interval. At each step of the VA, the phase at the start of the interval is calculated from the updated phase, frequency, and frequency rate estimates of the EKF of the previous interval. Define the EKF updated state estimates, at the mth interval for a possible bit edge, ϕ, as

$$x^{+}_{m,\phi,EKF} = \begin{bmatrix} \theta^{+}_{m,\phi,EKF} \\ \omega^{+}_{m,\phi,EKF} \\ \alpha^{+}_{m,\phi,EKF} \end{bmatrix}$$

Define the EKF propagated state estimates, at the mth interval for a possible bit edge, ϕ, as

$$x^{-}_{m,\phi,EKF} = \begin{bmatrix} \theta^{-}_{m,\phi,EKF} \\ \omega^{-}_{m,\phi,EKF} \\ \alpha^{-}_{m,\phi,EKF} \end{bmatrix}$$

Define these parameters at the start of the interval as $\hat{\theta}_{e_m,\phi}$, $\hat{\omega}_{e_m,\phi}$, and $\hat{\alpha}_{e_m,\phi}$. The relationships between the EKF updated outputs at the mth interval and these values at the start of the $(m+1)$st interval can be found as follows. First, the relationships between the EKF updated outputs at the mth interval and these values at the start of the mth interval are

$$\theta^{+}_{m,\phi,EKF} = \hat{\theta}_{e_m,\phi} + \frac{1}{T_{d_m}} \int_0^{T_{d_m}} \hat{\omega}_{e_m,\phi}\, t\, dt + \frac{1}{T_{d_m}} \int_0^{T_{d_m}} \hat{\alpha}_{e_m,\phi} \frac{t^2}{2}\, dt \quad (4.61)$$

$$\theta^{+}_{m,\phi,EKF} = \hat{\theta}_{e_m,\phi} + \hat{\omega}_{e_m,\phi} \frac{T_{d_m}}{2} + \hat{\alpha}_{e_m,\phi} \frac{T^2_{d_m}}{6} \quad (4.62)$$

Similarly,

$$\omega^{+}_{m,\phi,EKF} = \hat{\omega}_{e_m,\phi} + \hat{\alpha}_{e_m,\phi} \frac{T_{d_m}}{2} \quad (4.63)$$

$$\alpha^{+}_{m,\phi,EKF} = \hat{\alpha}_{e_m,\phi} \quad (4.64)$$

The EKF propagates the average phase, frequency, and frequency rate over the $(m+1)$st interval as follows:

$$\theta^{-}_{m+1,\,\phi,EKF} = \theta^{+}_{m,\phi,EKF} + \omega^{+}_{m,\phi,EKF}\, T_{d_m} + \alpha^{+}_{m,\phi,EKF} \frac{T^2_{d_m}}{2} \quad (4.65)$$

$$\omega^{-}_{m+1,\,\phi,EKF} = \omega^{+}_{m,\phi,EKF} + \alpha^{+}_{m,\phi,EKF}\, T_{d_m} \quad (4.66)$$

$$\alpha^{-}_{m+1,\,\phi,EKF} = \alpha^{+}_{m,\phi,EKF} \quad (4.67)$$

The frequency rate, frequency, and phase at the start of the $(m+1)$st interval are

$$\hat{\alpha}_{e_{m+1},\phi} = \alpha^{-}_{m+1,\,\phi,EKF} \quad (4.68)$$

$$\hat{\omega}_{e_{m+1},\phi} = \omega^{-}_{m+1,\,\phi,EKF} - \hat{\alpha}_{e_{m+1},\phi} \frac{T_{d_m}}{2} \quad (4.69)$$

$$\hat{\theta}_{e_{m+1},\phi} = \theta^{-}_{m+1,\,\phi,EKF} - \hat{\omega}_{e_{m+1},\phi} \frac{T_{d_m}}{2} - \hat{\alpha}_{e_{m+1},\phi} \frac{T^2_{d_m}}{6} \quad (4.70)$$

These values are used to obtain the phase error estimate at each of the following 20 T_{c_i}-sec. signals of the $(m+1)$st interval. If n defines the index of each T_{c_i}-sec. signal with respect to the start of the interval, then the estimate at each of the T_{c_i}-sec. signals is

$$\hat{\theta}_{e_{m+1},\,n,\,\phi} = \hat{\theta}_{e_{m+1},\phi} + T_{m+1,n}\, \hat{\omega}_{e_{m+1},\phi} + \frac{1}{2} T^2_{m+1,n}\, \hat{\alpha}_{e_{m+1},\phi}$$
$$\text{for } n = 0, \ldots, 19 \quad (4.71)$$

where $T_{m+1,n} = T_{m+1,n-1} + T_{c_{m+1,n}}$. $T_{m+1,0} = 0$. $T_{c_{m+1,n}}$ is the Doppler-compensated code length of the nth signal within the $(m+1)$st interval. In the case of low dynamics, $T_{m+1,n} = n\, T_{c_{m+1,0}}$. The phase values in (4.71) are used to remove the phase error from each of the following 20 T_{c_i}-sec. signals by counterrotating these signals. The rotated signals are averaged and used to calculate the weight function of the VA. The new cumulative weights are obtained, and then the value of the current data bit is estimated, which in turn is used in the $(m+1)$st interval of the EKF calculations to remove the effect of the data bit sign.

The EKF operation is summarized as follows:

$$Res_{m,\phi} = \frac{1}{\sqrt{20}} \sum_{n=0}^{19} \Im \left\{ y_{m,n,\phi}\, e^{-j\, \hat{\theta}_{e_{m,n,\phi}}} \right\}$$

$$H_{m,\phi} = [\hat{d}_{m,\phi}\ \hat{A}_{m,\phi}\ \ 0\ \ 0]$$

where $y_{m,n,\phi}$ refers to the signal number $(20\,(m-1) + n + \phi)$ of (4.59); $\hat{d}_{m,\phi}$ is the estimated data bit value; and $\hat{A}_{m,\phi}$ is an estimate of the current signal level obtained as

$$\hat{A}_{m,\phi} = \frac{1}{\sqrt{20}} \left| \sum_{n=0}^{19} \left\{ y_{m,n,\phi}\, e^{-j\, \hat{\theta}_{e_{m,n,\phi}}} \right\} \right|$$

The EKF equations in [7, 8] are applied as follows:

$$P^{-}_{m,\phi} = \Phi\, P^{+}_{m-1,\phi}\, \Phi^{T} + Q_{n_m}$$

$$K_{m,\phi} = P^{-}_{m,\phi}\, H^{T}_{m,\phi} \left(H_{m,\phi}\, P^{-}_{m,\phi}\, H^{T}_{m,\phi} + R_n \right)^{-1}$$

$$P^{+}_{m,\phi} = (I - K_{m,\phi}\, H_{m,\phi})\, P^{-}_{m,\phi}$$

where $K_{m,\phi}$ is the gain at the mth interval for a possible bit edge ϕ; $P^{-}_{m,\phi}$ is the propagated error covariance; and $P^{+}_{m,\phi}$ is the updated error covariance. The state estimate is then updated as

$$x^{+}_{m,\phi,EKF} = x^{-}_{m,\phi,EKF} + K_{m,\phi}\, Res_{m,\phi}$$

If the EKF output is used to update the local signal generator, then there will be no need to counterrotate the T_{c_i}-sec. y_i signals by any phase value. Therefore, (4.71) will not be calculated. The residual is found from

$$Res_{m,\phi} = \frac{1}{\sqrt{20}} \sum_{n=0}^{19} \Im \{y_{m,n,\phi}\}$$

4.8 Use of the Repeated Subframes

The navigation message, described in Section 1.5, has some data that repeat every specified time. Words 3 to 10 of subframes 1 to 3, which contain the ephemeris, repeat every 30 seconds. Words 3 to 10 of subframes 4 and 5, which contain the almanac, repeat every 12.5 minutes. This message structure is used in detecting the data of very weak signals.

The data detection limit, or the optimal BER, is found from the theoretical BER of the BPSK, which is

$$\text{BER} = \frac{1}{2}\ \text{erfc}(\sqrt{\text{SNR}}) \tag{4.72}$$

where erfc is the complementary error function, and SNR is the signal-to-noise ratio. The SNR is related to the C/N_0 by the coherent integration interval T_I as follows:

$$\text{SNR} = C/N_0\ T_I \tag{4.73}$$

So,

$$\text{BER} = \frac{1}{2}\ \text{erfc}(\sqrt{C/N_0\ T_I}) \tag{4.74}$$

The BER for the data sequence estimated by the VA approaches the theoretical limit, but this limit is high at very-low-power signals (e.g., BER = 0.13 for $C/N_0 =$ 15 dB-Hz).

Since the raw data bits located at indices 61 to 300 of subframes 1, 2, and 3 repeat every 30 seconds, then there are three sequences that repeat every 1,500 data bits. Each sequence contains 240 data bits. Each of the T_{c_i}-sec. correlated signals will have the same data content every $1,500 \times 20$ correlated signals. Since the noise is AWGN with zero mean, the coherent averaging of signals with the same data content will result in an increase in the SNR. This is used to achieve lower BER and higher EDR, as compared to the case when no coherent averaging of the repeated data is done. The repeated data are used as follows. The observations $s_{m,\phi}$

are generated as in (4.10) or (4.56). The observations that belong to a repeated sequence, have the same ϕ value, and have their indices m that generate the same value from (m mod 1, 500) are averaged together coherently to get new observations as

$$sr_{m,\phi} = \frac{1}{M} \sum_{b=0}^{M-1} s_{m+(b*1500),\,\phi} \tag{4.75}$$

where M is the number of repeated sequences with the same content that are averaged together. $sr_{m,\phi}$ replaces $s_{m,\phi}$ in (4.16) to (4.20); then, the VA proceeds with its estimation.

If the correlated signals have phase, Doppler shift, or Doppler rate errors, then these errors are estimated and removed prior to adding the signals of the repeated data. This is because each repeated sequence will have different values of these errors. There are two possible implementations of the EKF, as discussed in Section 4.7.1. If the EKF tracks the errors, then the estimated errors will be used to counterrotate the signals. Following that, the signals of the repeated data are averaged as in (4.75). Then, the algorithm proceeds to find the new cumulative sum and the current data bit estimate.

If the algorithm operates on stored data, then it is possible to utilize the repeated subframes in both the bit edge and the data sequence estimation. In this case, each sequence that has the same data content can operate a separate EKF. After the counterrotation and forming of $s_{j,\,\phi}$, signals of each repeated sequence are averaged as in (4.75). Then, the algorithm proceeds to find the new cumulative summation and the current data bit estimate at each ϕ, where all the EKFs of the same repeated sequence and the same possible bit edge ϕ use the last data bit estimate of this ϕ to remove the effect of the data bit sign from the next EKF estimates.

The decrease in the BER is found by calculating the increase in the SNR. Adding M data bits that have the same value, each of which is corrupted by AWGN, is equivalent to an increase in the data bit length by a factor M. So,

$$\text{SNR}_{new} = C/N_0 \sum_{b=1}^{M} T_{d_b} \tag{4.76}$$

where T_{d_b} is a Doppler-compensated data bit length. This is equivalent to an increase in the SNR of $10 \log M$ dB (e.g., if $M = 2$, the SNR increases by 3 dB). The new BER becomes

$$\text{BER}_{new} = \frac{1}{2}\ \text{erfc}\left(\sqrt{C/N_0 \sum_{b=1}^{M} T_{d_b}}\right) \tag{4.77}$$

4.9 Computational Analysis and Reduction

The number of computations, or the algorithm complexity, is derived as follows. A separate state diagram is used for each possible bit edge position because the bit edge is constant over the whole sequence. Define v as the number of possible data bit values ($v = 2$) and g as the number of possible bit edge positions ($g = 20$). The computation of the $s_{m,\phi}$ values is done as follows. At the first step for $\phi = 1$, $s_{1,1}$ is computed by averaging the first 20 signals. Then, the remaining values of $s_{1,\phi}, \phi = 2, \ldots, 20$, are found by a method similar to the one adopted in the ML estimation in [9], which can be written as

$$s_{1,\phi} = s_{1,\phi-1} - y_{\phi-1} + y_{\phi+19} \tag{4.78}$$

In the remaining steps for $m = 2, \ldots, N$, $s_{m,1}$ is found from

$$s_{m,1} = s_{m-1,20} - y_{(m-1)*20} + y_{m*20} \tag{4.79}$$

The remaining values of $s_{m,\phi}$, for $m = 2, \ldots, N$ and $\phi = 2, \ldots, 20$, are found from

$$s_{m,\phi} = s_{m,\phi-1} - y_{(m-1)*20+\phi-1} + y_{m*20+\phi-1} \tag{4.80}$$

The total amount of computation needed to calculate $s_{m,\phi}$ is

$$C_1 = g + 2(g-1) + \sum_{m=2}^{N} 2g = 2gN + g - 2 \tag{4.81}$$

At each step, there are two possible values for a data bit $(+1, -1)$ and 20 possible values for the bit edge position. There are $v * g = 2 * 20$ possible values for $|s_{m,\phi} - \bar{A}\, d_m|^2$. Each value is calculated by a subtraction and a multiplication. For the N steps, the complexity is

$$C_2 = 2vgN \tag{4.82}$$

The number of comparisons needed to find $\Gamma_{m,\phi}$ at each step is $v - 1$ for each possible bit edge position. For N steps, the complexity is

$$C_3 = (v-1)\, gN \tag{4.83}$$

At the last step, the minimum value of all the $\Gamma_{N,\phi}$ is found from the g surviving paths. Each survivor path contains estimates of the data sequence and corresponds to a possible bit edge position. The minimum of all $\Gamma_{N,\phi}$ is found using one of the sorting algorithms in [1]. This is done by recursively dividing all the values into two groups until each group has one value. For example, if there are 8 values, then these values are divided into 2 groups, each with 4 values, and so on.

Thus, there will be $\log_2 8$ levels of groups. Then, the minimum of each two groups at the lowest level is obtained by one comparison, and then each two groups at the same level are combined. This is repeated at all levels, starting from the last level. The number of comparisons needed to find the minimum of all $\Gamma_{N,\phi}$ is

$$\begin{aligned} C_4 &= \frac{g}{2} + \frac{g}{4} + \frac{g}{8} + \ldots + 2 + 1 \\ &= 2^{\log_2(g)-1} + 2^{\log_2(g)-2} + 2^{\log_2(g)-3} + \ldots + 2^1 + 2^0 \\ &= \sum_{i=0}^{\log_2(g)-1} 2^i = \frac{2^{\log_2(g)} - 1}{2-1} \\ &= g - 1 \end{aligned} \tag{4.84}$$

From (4.81) to (4.84), at each step m, except for the first and the last steps, the number of computations needed is

$$C_m = g + 3vg \tag{4.85}$$

The total algorithm complexity over N steps is

$$C_N = 3vgN + gN + 2g - 3 \tag{4.86}$$

By substituting $v = 2$ and $g = 20$, the following is obtained:

$$C_N = 140\,N + 37 \tag{4.87}$$

In the case of the repeated data, the added computation will be due to the averaging needed to generate the $s\,r_{m,\phi}$ values from the M sequences. Using similar analysis the total complexity is

$$C_N = 3vgN + MgN + 2g - 3 \tag{4.88}$$

By substituting $v = 2$ and $g = 20$ into (4.88), the following is obtained:

$$C_N = 120\,N + 20\,M\,N + 37 \tag{4.89}$$

The number of computations needed when using an EKF is calculated as follows. Define C_{EKF} as the number of computations needed at each step of the EKF operation. Since each possible bit edge has a separate EKF, the total number of computations of all the EKFs is $g\,C_{EKF}$. At each step m, the amount of phase rotation is found from the EKF output as in (4.68) to (4.71); this operation needs $O(g^2)$ computations. The counterrotations of the next 20 signals for all the ϕ's need a total of $O(g^2)$ computations. Forming the average values from the counterrotated signals requires a total of $O(g^2)$ computations. If finding the absolute value of the resultant I and Q signals requires C_{abs} computations, then a total of $g\ C_{abs}$ computations are needed for all the ϕ's. Adding all the previous computations and multiplying by the number of steps, N, result in a complexity of

$$N\left(g\ C_{EKF} + g\ C_{abs} + O(g^2)\right)$$

The algorithm then proceeds as when there are no phase, Doppler shift, or Doppler rate errors. Therefore, the total number of computations is

$$C_N = N\left[g\ C_{EKF} + O(g^2) + 3vg + (C_{abs} - 1)\ g\right] + g - 1 \tag{4.90}$$

This indicates that the total complexity of the algorithm is bounded by $O(N(g^2 + g\ C_{EKF}))$; the complexity needed in each step is bounded by $O(g^2 + g\ C_{EKF})$.

The number of computations can be reduced by discontinuing tracking of some of the ϕ's after a number of bit intervals of $N_{disBits}$. This is justified since the cumulative sum should largely increase when the tested bit edges are located farther away from the correct one as compared to the cumulative sum at the correct bit edge, and the cumulative sum should moderately increase when the tested bit edges are located near to the correct one as compared to the cumulative sum at the correct bit edge. Also, the difference between the maximum cumulative sum and the minimum one increases with time. So, the bit edges that seem most likely not to be the correct ones do not need to be tracked throughout all N steps. This is done by discarding a number of possible bit edges, $N_{disEdge_\zeta}$, after each number of bit intervals $N_{disBits_\zeta}$; the discarded possible bit edges are those with the maximum cumulative sum. The process is done several times until only one possible bit edge remains. The number of discarded edges in each process is specified in the vector $N_{disEdge}$, while the number of the elapsed steps before each discarding process is specified in the vector $N_{disBits}$. This leads to a reduction of $O(\sum_{i=1}^{\zeta} N_{disEdge_i}^2 + \sum_{i=1}^{\zeta} N_{disEdge_i}\ C_{EKF})$ in the computations needed at each step after the ζth discarding process.

4.10 Simulation and Results

The developed algorithms are demonstrated using simulated GPS signals with C/A codes. Settings similar to those used in Section 3.7 are used here, except that the sampling rate f_s is 5, 700 kHz. In addition, the initial phase error, Doppler shift, Doppler shift error, Doppler rate, and Doppler rate error are modeled as uniformly distributed random variables in the ranges of $(-\pi,\ \pi)$ radians, $(-10,\ 10)$ kHz, $(-200,\ 200)$ Hz, $(-300,\ 300)$ Hz/s, and $(-20,\ 20)$ Hz/s, respectively. The use of the Doppler shift error range as $(-200,\ 200)$ Hz corresponds to a coherent integration of at most 3 ms in the acquisition process. A larger coherent integration is used in the acquisition algorithms developed in Chapter 3 that will result in a much smaller Doppler shift error. However, those values have been used to demonstrate the performance of the algorithm in general; the same applies for the assumed Doppler rate error range. The first received bit edge is modeled as a uniformly distributed random variable in the range of (1, 20). Two clocks are simulated: temperature-compensated crystal oscillator (TCXO) and the ovenized crystal oscillator (OXO). The clock model is the same as that described in Section 3.7. The values of the phase and frequency random-walk intensities for the TCXO clock are $S_f = 5 \times 10^{-21}$ s and $S_g = 5.9 \times 10^{-20}\ s^{-1}$, and those values for the OXO clock are $S_f = 5 \times 10^{-23}$ s and $S_g = 1.5 \times 10^{-22}\ s^{-1}$.

Performance evaluations are done for the ES-VA for fine acquisition, the VA in the case of no carrier-tracking errors, the VA using the repeated subframes method to reduce the BER with a number of repeated bits of 2, 5, and 10, and the VA with the EKF in the case of phase, Doppler shift, and Doppler rate errors. In the cases of VA without carrier-tracking errors and the VA with the repeated subframes, 200 bits (i.e., 4 seconds of data) are used. For the VA case with the EKF, 400 bits are used (i.e., 8 seconds of data). In all cases, the C/N_0 ranges from 15 to 24 dB-Hz. To get accurate statistics, 10,000 trials are performed for the VA with no carrier-tracking errors and the VA with the repeated subframes, while 4,000 trials are performed for the VA with the EKF.

The ES-VA is used to initialize the EKF phase, frequency, and frequency rate. The settings for the ES-VA for the frequency initialization are $N_{freq} = 6$, $S_{separation} = 2$ ms, $R_{reduction} = 0.5$, and $f_{resolution} = 2$ Hz. The settings for the ES-VA for the phase initialization are: $N_{phase} = 12$ at the first iteration and $N_{phase} = 5$ at the rest, and $P_{resolution} = 5°$. Figure 4.4 shows the standard

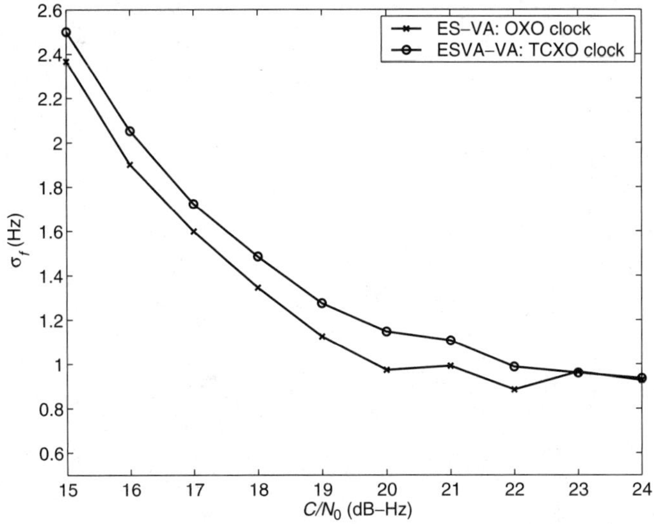

Figure 4.4 Standard deviation of the frequency estimation error of the ES-VA fine acquisition.

deviation of the frequency error estimation that results from the ES-VA fine acquisition algorithm; Figure 4.5 shows the standard deviation of the phase error estimation.

The fine acquisition of signals with high Doppler rate is tested for C/N_0 between 15 and 22 dB-Hz using a TCXO clock. The standard deviations of α_e and f_e are calculated. The Doppler rate error is estimated first using the ES-VA with the assumption that $f_e = 0$. Then, the frequency error is estimated using the frequency and rate estimate (FRE) method described in Section 4.6. A Doppler rate error is calculated for each tested frequency error using (4.47); then, the ES-VA is applied to find estimations for both α_e and f_e. Four different estimation times are used for both T_f and T_α, which are 4, 8, 12, and 16 seconds (i.e., 200, 400, 600, and 800 data bits). Figures 4.6 and 4.7 show the standard deviations of the frequency estimation error, σ_{f_e}, and the Doppler rate estimation error, σ_{α_e}, respectively. As shown in the figures, the algorithm is able to estimate α_e to an accuracy less than 1 Hz/s, while f_e estimation is not as accurate as α_e. However, the accuracy of α_e is more important than that of f_e, and these results can be used to initialize the EKF. These figures indicate that either 200 or 400 data bits are enough for the initialization process.

The EKF initial error covariance matrix P_1^- is set as a 3×3 diagonal matrix, with the three diagonal values equal to $P^2_{resolution}$, $(2\pi f_{resolution})^2$, and ϵ^2, where ϵ is a small value. The process noise values Q_θ, Q_ω, and Q_{α_ω} are set based on the clock type. The process noise values can also be chosen empirically to allow the EKF to converge; after the convergence, they are set based on the clock type.

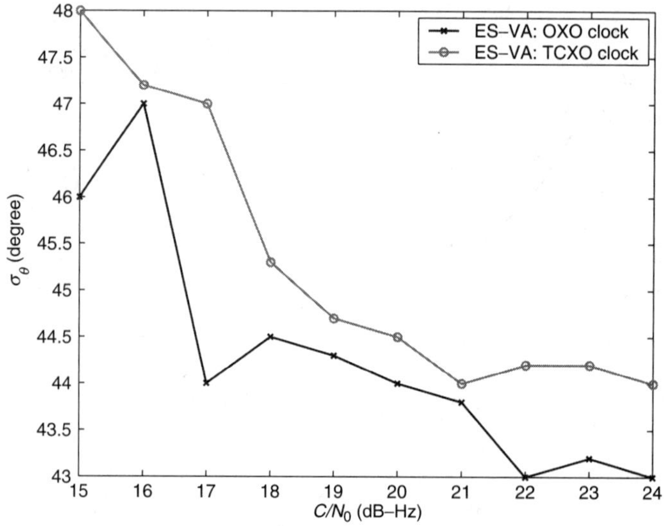

Figure 4.5 Standard deviation of the phase estimation error of the ES-VA fine acquisition.

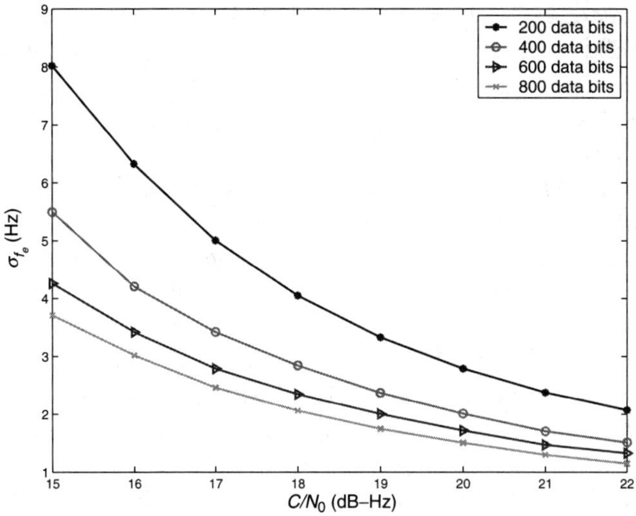

Figure 4.6 High dynamics: standard deviation of frequency error estimation using the ES-VA with a TCXO clock.

The BER and the EDR are directly related to the EKF phase and frequency estimation performance. If the frequency error is small (less than 0.5 Hz), then the BER and the EDR degrade slightly. As the frequency error increases, the BER degrades quickly because a frequency error causes a change in the data bit

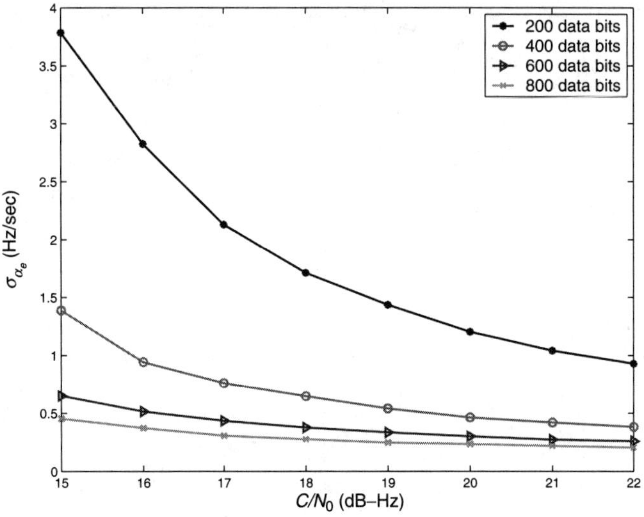

Figure 4.7 High dynamics: standard deviation of Doppler rate error estimation using the ES-VA with a TCXO clock.

polarity at random. The degradation in the EDR is related to the location and the number of the zero crossings caused by the frequency error, so it is not as bad for frequency errors less than about 5 Hz. However, for larger frequency errors, the EDR decreases and approaches zero at a frequency error of 25 Hz.

Two statistics are calculated to evaluate the performance of the EKF. The first one is the EKF frequency convergence rate (FCR). It is calculated as the percentage of the number of trials in which the average absolute frequency error over the last 2 seconds is less than 0.5 Hz. Errors in the frequency estimation degrade the BER and cause errors in the phase estimation. This can lead to cycle slips. Cycle slips can also occur even with frequency convergence; in this case, the cycle slips are related to the EKF phase estimation performance. A phase error of 180° causes the data bits to be detected with inverted signs. This degrades the overall BER, since such half-cycle slips are not detected. The second statistic, the convergence and tracking rate (CTR), is a measure of the overall performance of the EKF convergence and tracking capability throughout its operation. The CTR is calculated as the percentage of the number of trials that the EKF converges; correctly tracks the phase, frequency and frequency rate; and does not encounter any cycle slips. These statistics are shown in Figure 4.8. Also shown in the same figure is the percentage of correct frequency initialization (CFI) for the ES-VA. The CFI is 100% for all C/N_0 except at 15 dB-Hz when using a TCXO clock that encountered three cases, out of the 4,000 trials, of incorrect frequency initialization.

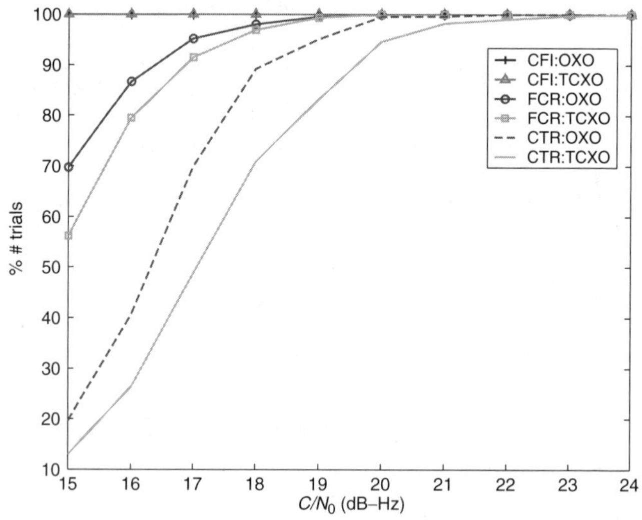

Figure 4.8 Percentage of the trials versus C/N_0, using OXO and TCXO clocks, for (a) CFI for ES-VA; (b) EKF FCR, and (c) EKF perfect CTR.

For comparison purposes, the histogram method for bit edge detection is simulated using thresholds of 570 and 600 (these values were used in [9]) and assuming no carrier-tracking errors. The bit edge detection is tested for the VA with no carrier-tracking errors and for the VA with the EKF (using the two clocks). The result is shown in Figure 4.9, in which the VA gives much better performance as compared to the performance of the histogram method. The VA with the EKF and the VA without carrier-tracking errors give similar performance for C/N_0 larger than 18 dB-Hz; otherwise, errors in the EKF phase and frequency estimation cause a small degradation in the EDR. Figure 4.10 shows the EDR and the standard deviation of the edge detection error (in milliseconds) over time for the VA with the EKF. The EDR is continuously increasing with time; it will reach one for all C/N_0. However, the amount of data needed to obtain an EDR of one depends on the C/N_0.

The following figures show the performance of only the VA when using an EKF. Figure 4.11 shows the BER over time for a total of 8 seconds. This BER is calculated for each 50 data bits, independently of the other data bits, to show the BER performance before and after the EKF convergence. Figure 4.12 is the total BER calculated for the last 7 seconds (i.e., 350 data bits) for the VA with the EKF. Also shown is the BER computed only for those cases in which the EKF converges and does not encounter any cycle slips. The theoretical BER and the BER of the VA without carrier-tracking errors are also shown; these two BERs are exactly the same. Figure 4.13 shows the BER of the VA when using the repeated subframes approach. This result indicates that it is possible to detect the data bits

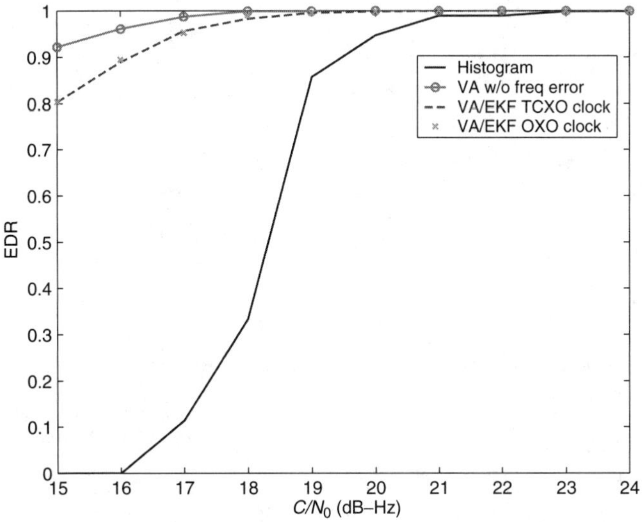

Figure 4.9 EDR versus C/N_0 of the histogram method, the VA without phase/frequency errors, and VA/EKF after 200 bits.

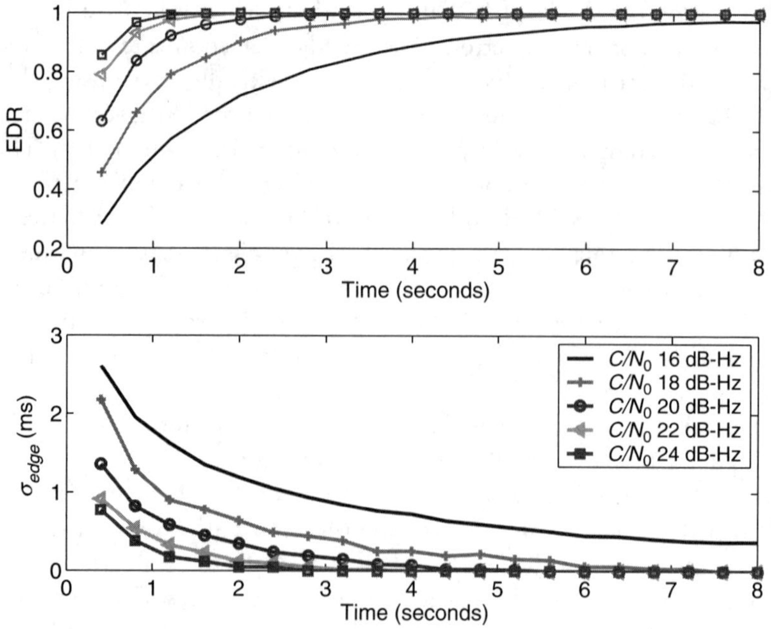

Figure 4.10 EDR and σ_{edge} of the detection error (ms) versus time of the VA/EKF for different C/N_0.

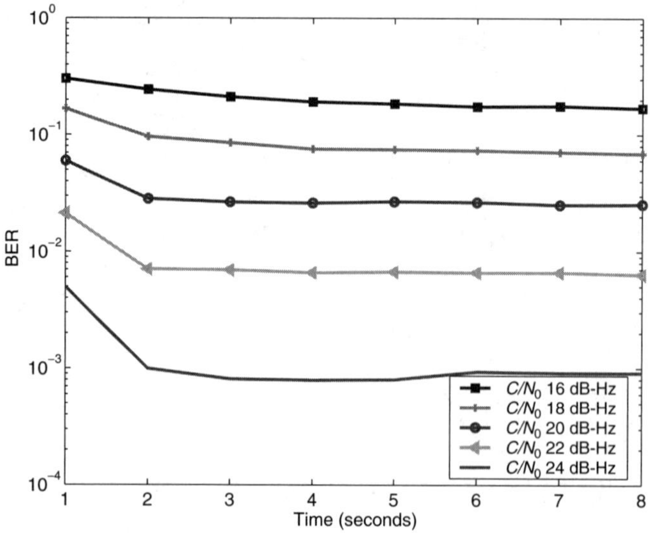

Figure 4.11 BER versus time of the VA/EKF for different C/N_0.

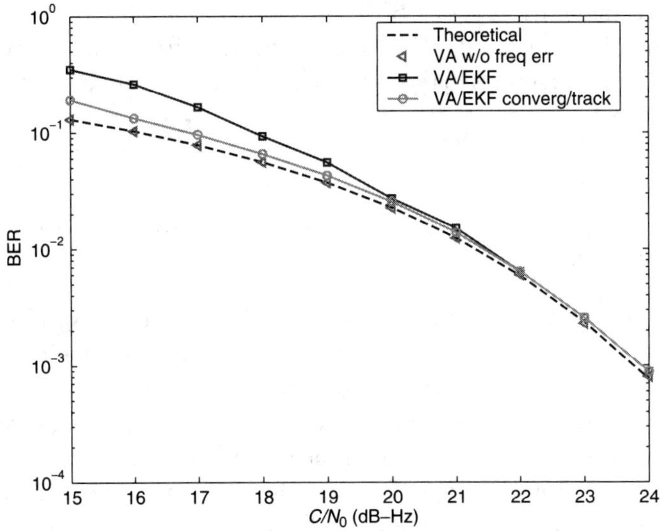

Figure 4.12 BER versus C/N_0 of the VA without phase/frequency errors, VA/EKF, and VA/EKF in the cases of the EKF convergence and perfect tracking; the theoretical BER is also shown.

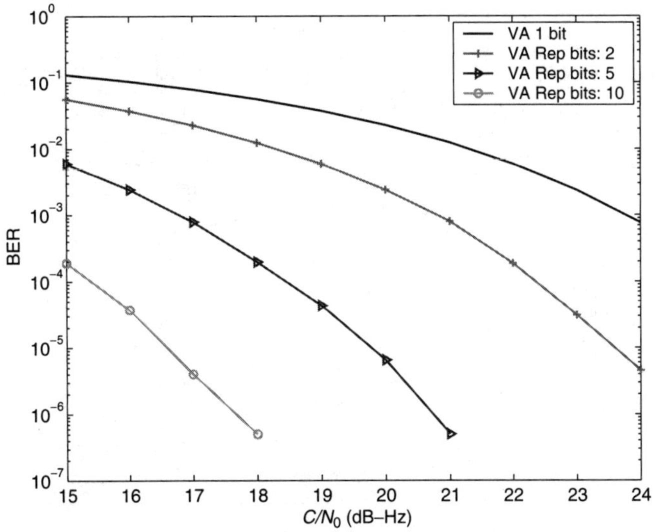

Figure 4.13 BER versus C/N_0 for the VA using repeated bits of 2, 5, and 10 compared to the VA with only 1 bit.

at any C/N_0 using such an approach. The number of the repeated subframes is chosen according to the required BER.

Figure 4.14 is the standard deviation of the EKF frequency error estimation over time. This corresponds to the variance of the errors of all the trials calculated at each time step as

$$\sigma_m^2 = \left(\frac{1}{N_{tr}} \sum_{n=1}^{N_{tr}} (f_{e_m,n} - \hat{f}_{e_m,n})^2 \right) - \left(\frac{1}{N_{tr}} \sum_{n=1}^{N_{tr}} (f_{e_m,n} - \hat{f}_{e_m,n}) \right)^2 \quad (4.91)$$

where σ_m^2 is the variance at the mth step of the EKF operation; N_{tr} is the number of trials; and $f_{e_m,n} - \hat{f}_{e_m,n}$ is the estimation error at the mth step for the nth trial.

Figure 4.15 is the standard deviation of the frequency error estimation for the EKF after 8 seconds of operation, using both the TCXO and OXO clocks. Figure 4.16 is the standard deviation of the phase error estimation for the EKF.

Figure 4.17 shows the time history of the EKF phase error estimation over all of the operation steps for one run of the algorithm for C/N_0 of 15, 18, and 24 dB-Hz. Figures 4.18(a) and 4.18(b) show the time history of the EKF frequency tracking, using TCXO clock and OXO clock, for C/N_0 of 15 dB-Hz and 18 dB-Hz, respectively. The true frequency error and the estimated one are shown in the upper part of the figure, while the estimation error is shown in the

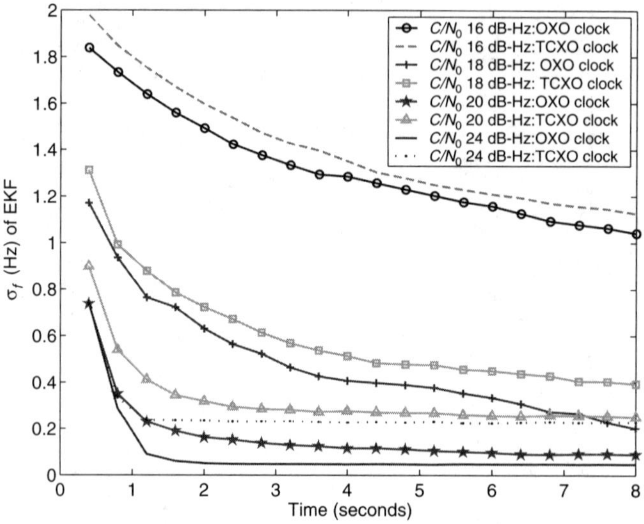

Figure 4.14 Standard deviation of the error in the EKF frequency estimation versus time using OXO and TCXO clocks for different C/N_0.

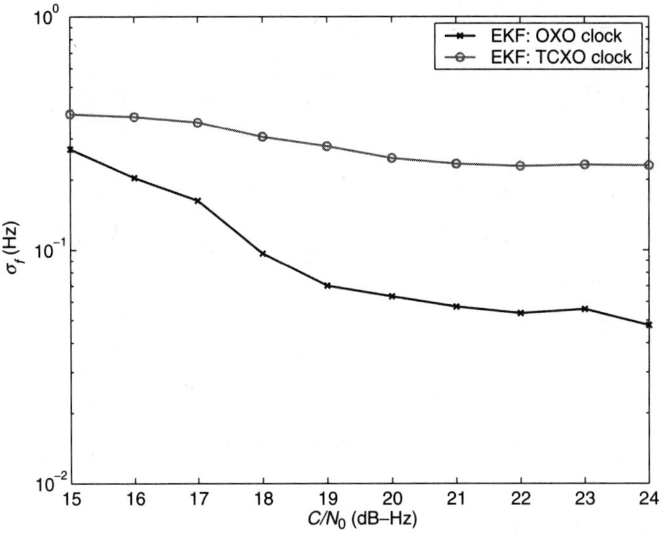

Figure 4.15 Standard deviation of the frequency estimation error of the EKF versus C/N_0 using OXO and TCXO clocks (log scale).

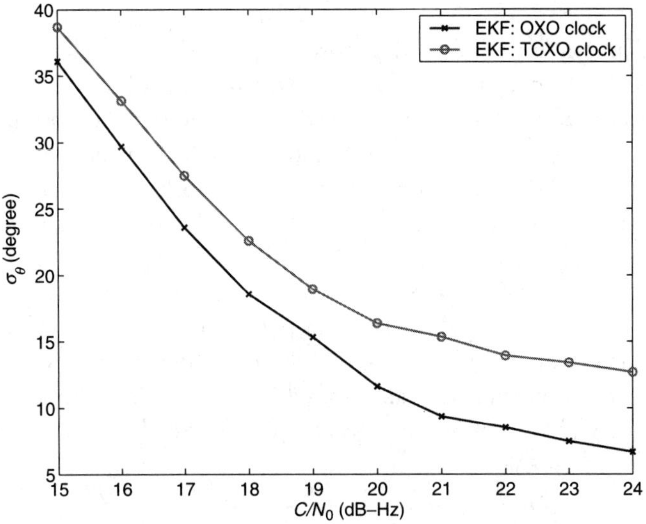

Figure 4.16 Standard deviation of the phase estimation error of the EKF versus C/N_0 using OXO and TCXO clocks (log scale).

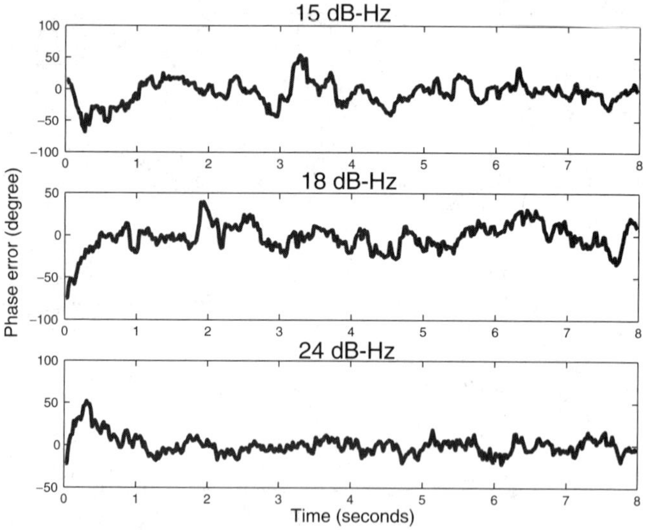

Figure 4.17 EKF phase estimation error history for C/N_0 15, 18, and 24 dB-Hz.

lower part. These results assume that the signal generator does not update its estimates after the fine acquisition.

4.11 Summary and Conclusions

A dynamic programming algorithm based on the VA is developed to estimate the bit edge position and to detect the navigation data bits at very low C/N_0 signals. A recursive ML function is derived to enable the application of the VA to this estimation problem. The ML function is used to associate a weight with each transition in the trellis graph. The algorithm works recursively to search for the path that most likely corresponds to the bit edge position and the data bit sequence. After each step, the algorithm keeps track of only one survivor path from each possible edge position. After any step, the survivor path with the minimum cumulative weight, among all the survivor paths up to this step, will contain the estimated data bits and correspond to the estimated bit edge position. This allows for access to the estimated data at any desired step. As the algorithm progresses, the estimated data will keep improving, reaching the optimal value at a step number depending on the C/N_0.

A fine acquisition algorithm based on the VA, ES-VA, is developed to enhance the carrier parameter estimation errors that result from the acquisition process. Also, a high dynamic fine acquisition algorithm is developed; it utilizes the relationship between the unknown Doppler shift and Doppler rate errors to avoid performing search over all the possible combinations of these errors.

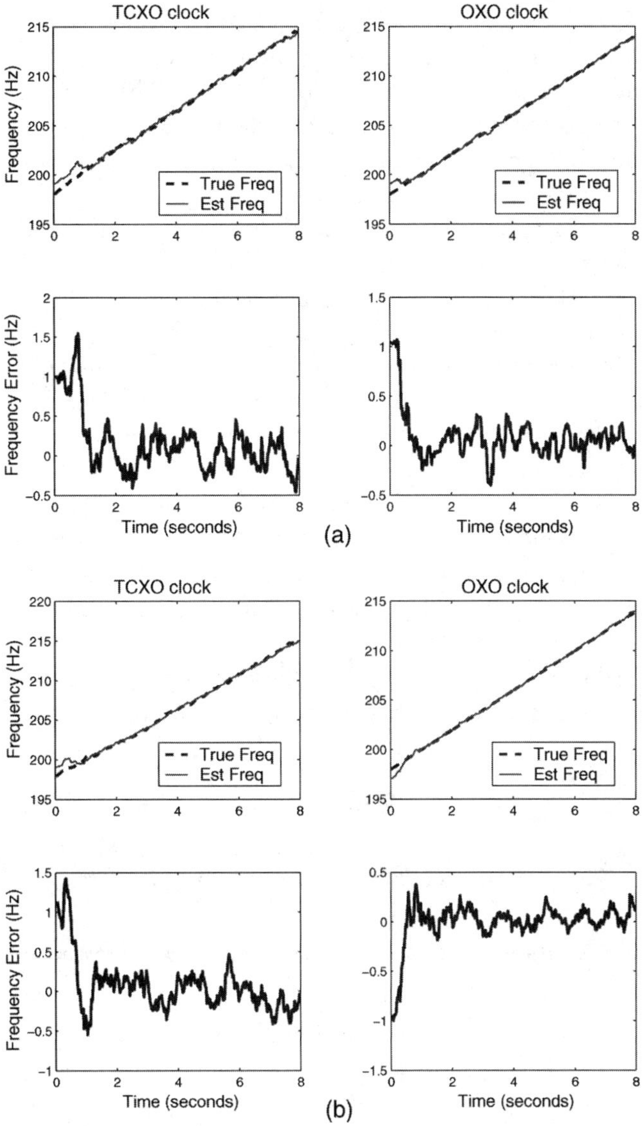

Figure 4.18 EKF frequency tracking history and estimation error history for OXO and TCXO clocks: (a) C/N_0 15 dB-Hz, and (b) C/N_0 18 dB-Hz.

Bit synchronization and data detection are accomplished in the presence of carrier-tracking errors by using EKFs in the VA paths to track these errors. EKF linearization is derived from the zero of the derivative of the log-likelihood of the received signal, given the last estimated errors of the phase, Doppler shift, and Doppler rate and the last estimated data bit value.

The BER is reduced by utilizing the repeated subframes. Signals that belong to repeated sequences and are separated by 30 seconds are added together; then, they are used to generate the weight function of the VA. This results in an increase in the SNR and, consequently, a decrease in the BER. The desired BER can be achieved at any C/N_0 by using an appropriate number of repeated subframes.

Simulation results indicate the ability of the algorithms to detect optimally the bit edge and the navigation data, even at very low C/N_0. The EKF gives high-quality convergence and tracking capabilities above C/N_0 of 18 dB-Hz for both the OXO and the TCXO clocks. The EKF sometimes has slow convergence below 18 dB-Hz. This causes a delay in obtaining a reliable estimate since the data cannot be detected reliably before the EKF converges. The BER is also sensitive to cycle slips. However, as long as any subframe is correctly synchronized, a half-cycle slip within that subframe will only lead to failure to decode the 30-bit word that encountered the slip. If all the data bits within one word, along with the last two bits of the preceding word, are detected with either correct or inverted signs, then this word will be correctly decoded, since the parity algorithm [10] is designed to detect and correct the inverted signs. In Chapter 5, an algorithm is developed to decode the navigation message at very low C/N_0; it deals with some practical issues associated with applying the repeated subframes approach. Edge detection is not sensitive to cycle slips and does not degrade badly with the slow convergence of the EKF. None of the developed algorithms need high computation time or high storage capacity.

References

[1] Cormen, T., C. Leisersen, and R. Rivest, *Introduction to Algorithms,* Boston MA: MIT Press, 1990.

[2] Parkinson, B., and J. Spilker, *Global Positioning System: Theory and Applications,* Washington, D.C.: AIAA, 1996.

[3] Mengali, U., and A. N. D'Andrea, *Synchronization Techniques for Digital Receivers,* New York: Plenum Press, 1997.

[4] Esteves, E. S., and R. Sampaio-Neto, "A Per-Survivor Phase Acquisition and Tracking Algorithm for Detection of TCM Signals with Phase Jitter and Frequency Error," *IEEE Trans Communications,* Vol. 45, November 1997, pp. 1381–1384.

[5] Vanelli-Coralli, A., et al., "A Performance Review of PSP for Joint Phase/Frequency and Data Estimation in Future Broadband Satellite Networks," *IEEE Trans Selected Areas in Communications,* Vol. 19, December 2001, pp. 2298–2309.

[6] Raheli, R., A. Polydoros, and C. Tzou, "Per-Survivor Processing: A General Approach to MLSE in Uncertain Environments," *IEEE Trans Communications,* Vol. 43, February 1995, pp. 354–364.

[7] Zarchan, P., and H. Musoff, *Fundamentals of Kalman Filtering: A Practical Approach,* Washington, D.C.: AIAA, 2000.

[8] Brown, R. G., and P. Y. C. Hwang, *Introduction to Random Signals and Applied Kalman Filtering,* New York: John Wiley & Sons, 1992.

[9] Kokkonen, M., and S. Pietila, "A New Bit Synchronization Method for a GPS Receiver," *Proc. IEEE Positioning, Location, and Navigation Symposium (PLANS)*, Palm Springs, CA, April 2002, pp. 85–90.

[10] NAVSTAR GPS Space Segment, Navigation User Interface Control Document (ICD-GPS-200), 1993–2000.

5

Code and Carrier Tracking and Navigation Message Decoding

5.1 Introduction

This chapter introduces code- and carrier-tracking algorithms designed to work with weak signals under various dynamic conditions. The code-tracking algorithm is based on EKF approaches. It implements a second-order EKF. Based on the signal condition and the correctly decoded data, it adaptively adjusts the coherent and incoherent integration times and the code-delay separation between the early and late signals. The carrier-tracking algorithm is also based on EKF approaches. It implements a square root EKF to circumvent numerical errors associated with the EKF. It also has the ability to adjust the integration time adaptively. Moreover, a navigation message decoding algorithm is developed. It utilizes the structure of the navigation message to enable decoding even for signals with a theoretical BER that does not allow for such decoding.

The performances of the code- and carrier-tracking and navigation message decoding algorithms are demonstrated for signals with carrier-to-noise ratios as low as 10-dB-Hz and up to 6-g acceleration. The results show the ability of the algorithms to work efficiently with such very weak signals

This chapter is organized as follows. Section 5.2 presents a description of the interaction between the code- and carrier-tracking modules. Section 5.3 presents the code-tracking module. Section 5.4 presents the carrier-tracking module. Section 5.5 presents the navigation message decoding algorithms. Section 5.6 presents some approaches designed to increase the time to lose lock for the tracking modules. Section 5.7 presents an approach to deal with a random, large, sudden changes in the receiver dynamics. Section 5.8 presents an overview of the computational requirements. Section 5.9 presents the simulation and results. Section 5.10 presents the summary and conclusions.

5.2 Tracking Module Interaction

The tracking consists of four modules: the signal generator, carrier tracking, code tracking, and navigation message decoding. Separate EKFs are used for the code and carrier tracking. Each EKF uses a different integration length. The carrier-tracking module is integrated with the VA, which detects the data bit values and provides an estimate of the signal level. The VA uses an integration length equal to the length of one data bit interval T_{d_m}. The EKFs of the code and carrier tracking use integration lengths that can be a multiple of one data bit interval. The integration length of the EKF used for the carrier tracking is defined as T_{car_u} at the uth integration interval. The integration length of the EKF used for the code tracking is defined as T_{code_v} at the vth integration interval. The lengths of T_{car_u} and T_{code_v} are not necessarily the same. T_{code_v} can be a combination of coherent and incoherent integrations; the total number of incoherent accumulations is defined as N_{incoh_v} at the vth interval. All the integration intervals start at the estimated beginning of the data bits.

The four modules use the output of each other. The outputs of the four modules are synchronized by having each module generate output every T_{d_m} sec. Thus, each EKF produces propagated estimates every T_{d_m} sec. The output of the carrier-tracking module consists of estimates of the phase, $\hat{\theta}_m$; phase error, $\hat{\theta}_{e_m}$; Doppler shift, $\hat{f}_{d_m}$; Doppler shift error, $\hat{f}_{e_m}$; Doppler rate, $\hat{\alpha}_m$; Doppler rate error, $\hat{\alpha}_{e_m}$; data bit value, $\hat{d}_m$; signal level, $\hat{A}_m$; and bit edge position. The output of the code-tracking module consists of estimates of the code delay, $\hat{\tau}_m$; code-delay error, $\hat{\tau}_{e_m}$; data bit length, taking into account the Doppler effect on the length T_{d_m}; and the start time of the next data bit, $t_{start_{m+1}}$. The signal generator output consists of the early, prompt, and late correlated signals. The navigation message decoding module generates two outputs: a flag F_{d_m} that is set to 1 to indicate that the mth data bit has been correctly decoded, while it is set to 0 otherwise; the other output is the correct data bit sign, defined as d_{L_m}. The estimates of the current code-delay error and the start of the next data bit are used by the signal generator to realign the replica code with the received one. This minimizes the code misalignment loss in the correlated signals. The estimates of the phase, Doppler shift, and Doppler rate are used by the local signal generator to minimize the correlation loss caused by errors in these parameters. Thus, the correlated signals are continuously generated with the current best estimates of the parameters. The code-tracking module uses the estimates of the Doppler shift and Doppler rate to estimate the code and data bit lengths, which are not constant because of the Doppler effect on those lengths. The EKF output of the carrier tracking is used by the EKF of the code tracking to generate the expected measurement and the linearization functions. The estimates of the code and data bit lengths are used by the EKF of the carrier tracking to generate the propagated estimates of the phase and Doppler shift. The interaction between the signal generator, code-tracking, and carrier-tracking modules is shown in Figure 5.1.

The C/N_0 of the received signal can be estimated at the beginning of the tracking. The expected BER, $\bar{\beta}$, is calculated from the estimated C/N_0. The actual BER, $\hat{\beta}$, is estimated over the repeated data bits that have previously been correctly decoded. This is done by comparing $\hat{d}_m$ to d_{L_m} over N_β data bits and counting the number of bits with incorrect signs N_{β_0}. Since there could be an undetected 180° phase error, $\hat{\beta} = \min\{N_{\beta_0}/N_\beta, (N_\beta - N_{\beta_0})/N_\beta\}$. If $N_{\beta_0}/N_\beta > (N_\beta - N_{\beta_0})/N_\beta$, then a flag F_{slip} is set to –1 to indicate the existence of a 180° phase error; otherwise F_{slip} is set to 1.

5.3 Code Tracking

The code tracking uses an integration time that is a multiple of one data bit interval. The total integration length for the vth interval is T_{code_v} sec. The integration can consist of both coherent and incoherent integrations. The number of incoherent integrations in the vth interval is defined as N_{incoh_v}. This means that N_{incoh_v} coherent integrations are added together to form the total integration. The lengths of the coherent integrations in the same integration interval do not have to be

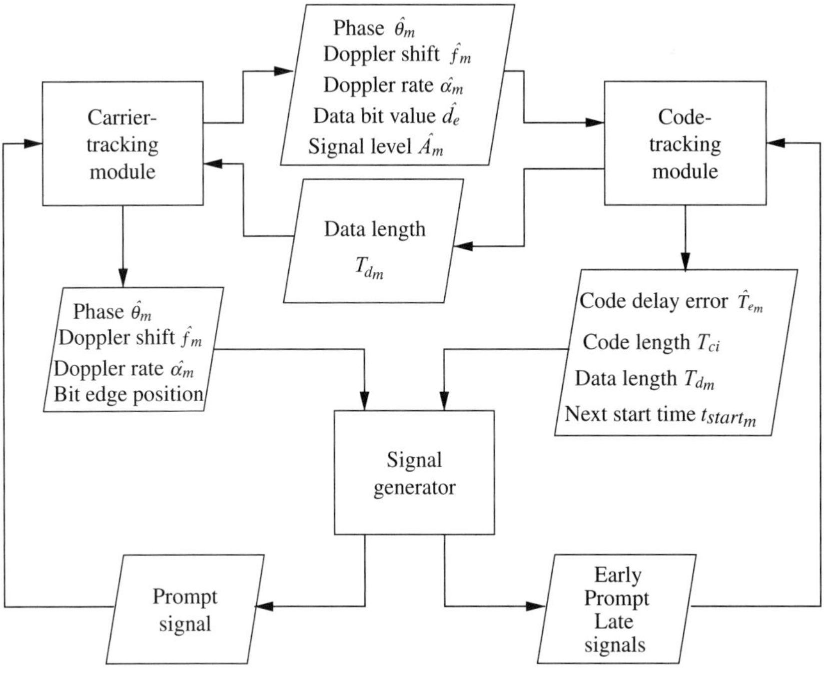

Figure 5.1 The interaction between the signal generator, carrier-tracking, and code-tracking modules.

equal. Each coherent integration interval can have a length of one data bit interval or a multiple of one data bit interval.

The EKF of the code-tracking module updates its estimates every T_{code_v} sec. However, the prompt, early, and late correlated signals are generated every T_{d_m} sec., where $T_{code_v} \geq T_{d_m}$. Therefore, the code-tracking module stores the correlated signals between the EKF update steps in a vector V_{EPL}. It also stores the carrier-tracking outputs that are needed in the code tracking. Those outputs include the following: (1) the estimated signal levels, which are stored in vector V_A; (2) the estimated phase errors, which are taken from the EKF update output and stored in a vector V_θ; and (3) the estimated data bit values, which are stored in vector V_d.

5.3.1 Timing Reconstruction and Local Signal Generation

The signal model described in Section 2.7 is used here. The objective of the code-tracking module is to align the local replica code with the received code. This is done by generating the samples of the local code at the same points as the samples of the received code. The first sample of the received code does not have to be located exactly at the start of the code. Since the module generates output every T_{d_m} sec., it estimates the time of the start of the mth interval, t_{start_m}, and uses it to calculate the time of the first sample of the mth interval, t_{first_m}. It should be noted that t_{start_m} and t_{first_m} also represent the estimated times of the start of the first code in the mth interval and the first sample of the mth data bit, respectively. Both t_{start_m} and t_{first_m} are calculated relative to the time of the start of the tracking. The time of the start of the tracking represents the time at which initial estimates are calculated for the phase, Doppler shift, Doppler rate, and code delay. This time is assumed here to be the start of the first data bit used in the tracking modules.

The t_{s_m} is the difference between the estimated times of the first sample in the mth interval and the start of that interval. It is defined as in (2.35) to (2.37); it is calculated as

$$t_{s_m} = t_{first_m} - t_{start_m} \tag{5.1}$$

In each m interval, the first sample of the first replica code is generated at a shift equal to t_{s_m} relative to the start of the code. This is done as follows. Define the module operation (i.e., the remainder calculation) of two variables a and b as

$$\{a \mod b\}$$

If the sampling rate is f_s and the time of the first received sample of the first interval used in the tracking is $t_0 = t_{first_1}$, where $0 \leq t_0 < 1/f_s$, then the time of all the samples t_k of the received signal must satisfy

$$\{(t_k - t_0) \mod T_s\} = 0 \tag{5.2}$$

where $T_s = 1/f_s$. If t_{start_m} satisfies (5.2), that is,

$$\left\{ \left(t_{start_m} - t_0 \right) \mod T_s \right\} = 0 \tag{5.3}$$

then the first sample in the mth interval is located exactly at the start of the code. So,

$$t_{first_m} = t_{start_m} \tag{5.4}$$

$$t_{s_m} = 0 \tag{5.5}$$

If t_{start_m} does not satisfy (5.3), then

$$t_{first_m} = t_{start_m} + T_s - \left\{ \left(t_{start_m} - t_0 \right) \mod T_s \right\} \tag{5.6}$$

$$t_{s_m} = T_s - \left\{ \left(t_{start_m} - t_0 \right) \mod T_s \right\} \tag{5.7}$$

The rest of the samples are generated at times equal to

$$t_{ml} = t_{s_m} + l T_s$$

where $l = 0, \ldots, N_m - 1$; N_m is the number of samples in the mth interval. However, the code is sampled, taking into consideration the effect of the Doppler shift on the code length. Therefore, the sampling times of the code are calculated from

$$t_{ml,\, f_d} = t_{ml} \left(1 + \frac{\hat{f}_{d_m} + \hat{\alpha}_m \left(t_{s_m} + l T_s \right) / 2}{f_{L1}} \right)$$

The code-tracking module also estimates the data bit length from

$$T_{d_m} = T_d \, \frac{f_{L1}}{f_{L1} + \hat{f}_{d_m} + \hat{\alpha}_m \, T_d / 2} \tag{5.8}$$

where T_d is the actual data bit length, which is equal to 0.02 sec.; $\hat{f}_{d_m}$ is the estimated Doppler shift at the start of the mth interval; and $\hat{\alpha}_m$ is the estimated Doppler rate for the mth interval.

One stream of the Doppler-compensated replica code is generated and used to generate the prompt, early, and late correlated signals. The stream has a number of code intervals equal to the number of code intervals in T_{dm} sec. plus parts of two chips that will be used to generate the early and late signals. Thus, the generated stream for the mth interval has a length equal to $T_{d_m} + 2\,\Delta$, where Δ is equal to half the code delay separation between the early and late signals. Δ is chosen to be a multiple N_s of the sampling time, given that it does not exceed the length of half a code chip. So,

$$\Delta = \frac{N_s}{f_s} \tag{5.9}$$

The times of the first and the last samples of the mth interval of the Doppler-compensated stream of replica code are calculated, respectively, from

$$tStartStream_m = t_{s_m} - \Delta \tag{5.10}$$

$$tEndStream_m = T_{d_m} - \{(T_{d_m} - t_{s_m}) \mod T_s\} + \Delta \tag{5.11}$$

The start and end sampling times of the stream are passed to the code generator. They are used to generate the early, prompt, and late correlated signals. The start sampling time of the three signals for the mth interval are, respectively,

$$t_{SE_m} = tStartStream_m \tag{5.12}$$

$$t_{SP_m} = tStartStream_m + \frac{N_s}{f_s} \tag{5.13}$$

$$t_{SL_m} = tStartStream_m + 2\,\frac{N_s}{f_s} \tag{5.14}$$

The end sampling times of the three signals are, respectively,

$$t_{EE_m} = tEndStream_m - 2\,\frac{N_s}{f_s} \tag{5.15}$$

$$t_{EP_m} = tEndStream_m - \frac{N_s}{f_s} \tag{5.16}$$

$$t_{EL_m} = tEndStream_m \tag{5.17}$$

The three local signals are correlated with the received signal. The time of the first sample of the mth received interval is t_{first_m}; the time of the last sample is calculated from

$$t_{last_m} = t_{first_m} - t_{s_m} + t_{EP_m} \tag{5.18}$$

Figure 5.2 illustrates the different times for the mth interval.

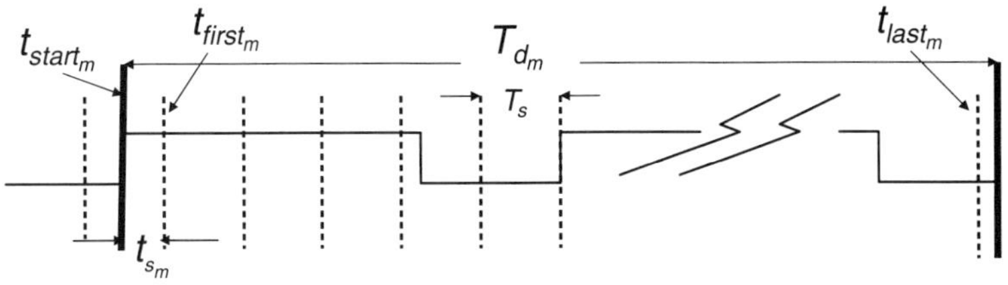

Figure 5.2 Illustration of different times for the *m*th interval.

If the bit edge position has not been estimated, then the prompt signal is generated with integration time equal to one code length, T_{c_i}. The T_{c_i}-sec. signals are used within the VA and the EKF, described in Chapter 4, Sections 4.4 to 4.7, to estimate the bit edge position, data bit sequence, phase, Doppler shift, and Doppler rate errors. If the bit edge position is identified, then the prompt signal is generated with integration time equal to one data bit interval, T_{d_m}. The early and late signals are generated with integration time equal to one data bit interval.

The in-phase and quad-phase parts of the correlated signals are

$$I_m(\delta) = \frac{1}{\sqrt{N_m}} \sum_{k=0}^{N_m-1} r\left(t_{first_m} + k/f_s\right) I_L\left(t_{s_m} + k/f_s + \delta\right) \tag{5.19}$$

$$Q_m(\delta) = \frac{1}{\sqrt{N_m}} \sum_{k=0}^{N_m-1} r\left(t_{first_m} + k/f_s\right) Q_L\left(t_{s_m} + k/f_s + \delta\right) \tag{5.20}$$

where δ is equal to $-\Delta$, 0, and Δ for the early, prompt, and late signals, respectively, and r, I_L and Q_L are as in (2.33), (2.35), and (2.36), respectively. The division by $\sqrt{N_m}$ is used to keep the noise variance equal to 1.

5.3.2 EKF Code Tracking

Let z_v and $\hat{z}_v$ define, respectively, the measurement vector and the expected measurement vector at the vth interval of the code tracking:

$$z_v = h_v + n_v \tag{5.21}$$

where h_v is a function in the unknown parameters that are to be estimated by the EKF, and n_v is the measurement noise.

Different EKF designs are developed, including first- and second-order EKFs. The first-order EKF uses the early and late signals, while the second-order uses also the prompt signals. For all the EKFs, the dynamics model has the following form

$$\tau_{e_{v+1}} = \Phi\, \tau_{e_v} + W_{\tau, v+1} \tag{5.22}$$

where W_τ is a noise term, $\Phi = 1$, and τ_{e_v} is the code-delay error.

To explain different EKF designs, z_v, $\hat{z}_v$, and the linearized H_v vectors are defined in general forms. These vectors need the in-phase and the quad-phase of the early, late, and prompt signals. The correlated signals are generated for each T_{d_m} sec. starting at the current estimated bit edge position.

If only a coherent integration is used, then the vectors are as follows. The general measurement vector is

$$z_v = \begin{bmatrix} I_{Ec_v} \\ Q_{Ec_v} \\ I_{Lc_v} \\ Q_{Lc_v} \\ I_{Pc_v} \\ Q_{Pc_v} \end{bmatrix} \tag{5.23}$$

where I_{Ec_v} and Q_{Ec_v} are the early in-phase and quad phase correlated signals; I_{Lc_v} and Q_{Lc_v} are the late correlated signals; and I_{Pc_v} and Q_{Pc_v} are the prompt correlated signals. z_v is obtained from

$$z_v = \begin{bmatrix} \sum_{m=g_v}^{g_v+N_v-1} \frac{1}{\sqrt{N_v}} \left[\left(1 - F_{d_m}\right) \hat{d}_m + F_{d_m} F_{slip_v} d_{L_m}\right] I_m(-\Delta) \\ \sum_{m=g_v}^{g_v+N_v-1} \frac{1}{\sqrt{N_v}} \left[\left(1 - F_{d_m}\right) \hat{d}_m + F_{d_m} F_{slip_v} d_{L_m}\right] Q_m(-\Delta) \\ \sum_{m=g_v}^{g_v+N_v-1} \frac{1}{\sqrt{N_v}} \left[\left(1 - F_{d_m}\right) \hat{d}_m + F_{d_m} F_{slip_v} d_{L_m}\right] I_m(\Delta) \\ \sum_{m=g_v}^{g_v+N_v-1} \frac{1}{\sqrt{N_v}} \left[\left(1 - F_{d_m}\right) \hat{d}_m + F_{d_m} F_{slip_v} d_{L_m}\right] Q_m(\Delta) \\ \sum_{m=g_v}^{g_v+N_v-1} \frac{1}{\sqrt{N_v}} \left[\left(1 - F_{d_m}\right) \hat{d}_m + F_{d_m} F_{slip_v} d_{L_m}\right] I_m(0) \\ \sum_{m=g_v}^{g_v+N_v-1} \frac{1}{\sqrt{N_v}} \left[\left(1 - F_{d_m}\right) \hat{d}_m + F_{d_m} F_{slip_v} d_{L_m}\right] Q_m(0) \end{bmatrix} \tag{5.24}$$

where g_v and N_v are, respectively, the index of the first data bit and the number of data bits in the vth interval. $\hat{d}_m$ is the estimated data bit value, which is taken from the carrier-tracking module output. F_{d_m} is taken from the output of the navigation message decoding module; it is set to 1 to indicate that the data bit has been correctly decoded, and it is set to 0 otherwise. d_{L_m} is the correct data bit value. F_{slip_v} is a flag that is set to −1 to indicate the existence of a 180° carrier phase error, and it is set to 1 otherwise. The division by $\sqrt{N_v}$ is done to keep the noise variance equal to 1. The general expected measurement vector is

$$\hat{z}_v = \begin{bmatrix} \sum_{m=g_v}^{g_v+N_v-1} \frac{1}{\sqrt{N_v}} \hat{A}_m \cos(\hat{\theta}_{e_m}) \hat{R}(-\Delta) \\ \sum_{m=g_v}^{g_v+N_v-1} \frac{1}{\sqrt{N_v}} \hat{A}_m \sin(\hat{\theta}_{e_m}) \hat{R}(-\Delta) \\ \sum_{m=g_v}^{g_v+N_v-1} \frac{1}{\sqrt{N_v}} \hat{A}_m \cos(\hat{\theta}_{e_m}) \hat{R}(\Delta) \\ \sum_{m=g_v}^{g_v+N_v-1} \frac{1}{\sqrt{N_v}} \hat{A}_m \sin(\hat{\theta}_{e_m}) \hat{R}(\Delta) \\ \sum_{m=g_v}^{g_v+N_v-1} \frac{1}{\sqrt{N_v}} \hat{A}_m \cos(\hat{\theta}_{e_m}) \hat{R}(0) \\ \sum_{m=g_v}^{g_v+N_v-1} \frac{1}{\sqrt{N_v}} \hat{A}_m \sin(\hat{\theta}_{e_m}) \hat{R}(0) \end{bmatrix} \tag{5.25}$$

where $\hat{A}$ and $\hat{\theta}_e$ are the estimated signal level and the carrier phase error at the middle point of each data bit, respectively; they are taken from the output of the carrier-tracking module. $\hat{R}(\Delta)$ is an approximation of the autocorrelation function, which depends on the EKF type. The general linearized vector is

$$H_{1v} = \frac{\partial h_v}{\partial \tau}\Big|_{\tau_{e_v}=\tau_{e_v}^-} = \begin{bmatrix} \sum_{m=g_v}^{g_v+N_v-1} \frac{1}{\sqrt{N_v}} \hat{A}_m \cos(\hat{\theta}_{e_m}) \frac{\partial \hat{R}(-\Delta+\tau_{e_v})}{\partial \tau}\Big|_{\tau_{e_v}=\tau_{e_v}^-} \\ \sum_{m=g_v}^{g_v+N_v-1} \frac{1}{\sqrt{N_v}} \hat{A}_m \sin(\hat{\theta}_{e_m}) \frac{\partial \hat{R}(-\Delta+\tau_{e_v})}{\partial \tau}\Big|_{\tau_{e_v}=\tau_{e_v}^-} \\ \sum_{m=g_v}^{g_v+N_v-1} \frac{1}{\sqrt{N_v}} \hat{A}_m \cos(\hat{\theta}_{e_m}) \frac{\partial \hat{R}(\Delta+\tau_{e_v})}{\partial \tau}\Big|_{\tau_{e_v}=\tau_{e_v}^-} \\ \sum_{m=g_v}^{g_v+N_v-1} \frac{1}{\sqrt{N_v}} \hat{A}_m \sin(\hat{\theta}_{e_m}) \frac{\partial \hat{R}(\Delta+\tau_{e_v})}{\partial \tau}\Big|_{\tau_{e_v}=\tau_{e_v}^-} \\ \sum_{m=g_v}^{g_v+N_v-1} \frac{1}{\sqrt{N_v}} \hat{A}_m \cos(\hat{\theta}_{e_m}) \frac{\partial \hat{R}(\tau_{e_v})}{\partial \tau}\Big|_{\tau_{e_v}=\tau_{e_v}^-} \\ \sum_{m=g_v}^{g_v+N_v-1} \frac{1}{\sqrt{N_v}} \hat{A}_m \sin(\hat{\theta}_{e_m}) \frac{\partial \hat{R}(\tau_{e_v})}{\partial \tau}\Big|_{\tau_{e_v}=\tau_{e_v}^-} \end{bmatrix} \tag{5.26}$$

where $\tau_{e_v}^-$ is the propagated code-delay error; it is equal to 0 because the algorithm keeps track of the estimated start of the code, so the propagated error is 0.

The second derivative of h_v, defined as H_{2v}, produces a number of matrices equal to the number of the measurements; the jth matrix is

$$H_{2v,j} = \frac{\partial^2 h_{v_j}}{\partial \tau^2}\Big|_{\tau_{e_v}=\tau_{e_v}^-} \tag{5.27}$$

where h_{v_j} is the jth row of h_v. H_{2v} is needed only in a second-order EKF.

If both coherent and incoherent integrations are used, then (5.23) to (5.26) are modified to reflect the use of the incoherent accumulation over N_{incoh_v} coherent integrations. Each coherent integration interval can have a different length. The measurement vector is expressed as

$$z_v = \begin{bmatrix} I_{Einc_v} \\ Q_{Einc_v} \\ I_{Linc_v} \\ Q_{Linc_v} \\ I_{Pinc_v} \\ Q_{Pinc_v} \end{bmatrix} \tag{5.28}$$

This vector is obtained from

$$z_v = \frac{1}{\sqrt{2\,N_{incoh_v}}} \begin{bmatrix} \sum_{k=1}^{N_{incoh_v}} \left\{ \sum_{m=g_{vk}}^{g_{vk}+N_{vk}-1} [(1-F_{d_m})\hat{d}_m + F_{d_m} F_{slip_k} d_{L_m}] \frac{I_m(-\Delta)}{\sqrt{N_{vk}}} \right\}^2 \\ \sum_{k=1}^{N_{incoh_v}} \left\{ \sum_{m=g_{vk}}^{g_{vk}+N_{vk}-1} [(1-F_{d_m})\hat{d}_m + F_{d_m} F_{slip_k} d_{L_m}] \frac{Q_m(-\Delta)}{\sqrt{N_{vk}}} \right\}^2 \\ \sum_{k=1}^{N_{incoh_v}} \left\{ \sum_{m=g_{vk}}^{g_{vk}+N_{vk}-1} [(1-F_{d_m})\hat{d}_m + F_{d_m} F_{slip_k} d_{L_m}] \frac{I_m(\Delta)}{\sqrt{N_{vk}}} \right\}^2 \\ \sum_{k=1}^{N_{incoh_v}} \left\{ \sum_{m=g_{vk}}^{g_{vk}+N_{vk}-1} [(1-F_{d_m})\hat{d}_m + F_{d_m} F_{slip_k} d_{L_m}] \frac{Q_m(\Delta)}{\sqrt{N_{vk}}} \right\}^2 \\ \sum_{k=1}^{N_{incoh_v}} \left\{ \sum_{m=g_{vk}}^{g_{vk}+N_{vk}-1} [(1-F_{d_m})\hat{d}_m + F_{d_m} F_{slip_k} d_{L_m}] \frac{I_m(0)}{\sqrt{N_{vk}}} \right\}^2 \\ \sum_{k=1}^{N_{incoh_v}} \left\{ \sum_{m=g_{vk}}^{g_{vk}+N_{vk}-1} [(1-F_{d_m})\hat{d}_m + F_{d_m} F_{slip_k} d_{L_m}] \frac{Q_m(0)}{\sqrt{N_{vk}}} \right\}^2 \end{bmatrix} \tag{5.29}$$

where g_{vk} and N_{vk} are, respectively, the index of the first data bit and the number of the data bit intervals in the kth coherent integration of the vth integration interval. The division by $\sqrt{N_{vk}}$ is done to keep the noise variance resulting from the coherent addition equal to 1, while the division by $\sqrt{2\,N_{incoh_v}}$ is done to keep the total noise variance equal to 1 after the incoherent addition. The division by $\sqrt{2\ N_{incoh_v}}$ is concluded as follows. The PDF resulting from the square of an AWGN has the form

$$f_{\mathbf{N}}(n) = \frac{1}{\sqrt{2}}\ e^{-\frac{n}{\sqrt{2}}} \tag{5.30}$$

This exponential function has a variance of 2. Adding N_{incoh_v} of such functions increases the noise variance to $2\ N_{incoh_v}$. This follows from Section 3.6.3. The general expected measurement vector is

$$\hat{z}_v = \frac{1}{\sqrt{2\ N_{incoh_v}}} \begin{bmatrix} \sum_{k=1}^{N_{incoh_v}} \left\{ \sum_{m=g_{vk}}^{g_{vk}+N_{vk}-1} \frac{1}{\sqrt{N_{vk}}} \hat{A}_m \cos\left(\hat{\theta}_{e_m}\right) \hat{R}(-\Delta) \right\}^2 \\ \sum_{k=1}^{N_{incoh_v}} \left\{ \sum_{m=g_{vk}}^{g_{vk}+N_{vk}-1} \frac{1}{\sqrt{N_{vk}}} \hat{A}_m \sin\left(\hat{\theta}_{e_m}\right) \hat{R}(-\Delta) \right\}^2 \\ \sum_{k=1}^{N_{incoh_v}} \left\{ \sum_{m=g_{vk}}^{g_{vk}+N_{vk}-1} \frac{1}{\sqrt{N_{vk}}} \hat{A}_m \cos\left(\hat{\theta}_{e_m}\right) \hat{R}(\Delta) \right\}^2 \\ \sum_{k=1}^{N_{incoh_v}} \left\{ \sum_{m=g_{vk}}^{g_{vk}+N_{vk}-1} \frac{1}{\sqrt{N_{vk}}} \hat{A}_m \sin\left(\hat{\theta}_{e_m}\right) \hat{R}(\Delta) \right\}^2 \\ \sum_{k=1}^{N_{incoh_v}} \left\{ \sum_{m=g_{vk}}^{g_{vk}+N_{vk}-1} \frac{1}{\sqrt{N_{vk}}} \hat{A}_m \cos\left(\hat{\theta}_{e_m}\right) \hat{R}(0) \right\}^2 \\ \sum_{k=1}^{N_{incoh_v}} \left\{ \sum_{m=g_{vk}}^{g_{vk}+N_{vk}-1} \frac{1}{\sqrt{N_{vk}}} \hat{A}_m \sin\left(\hat{\theta}_{e_m}\right) \hat{R}(0) \right\}^2 \end{bmatrix} \tag{5.31}$$

The general linearized vector is

$$H_{1v} = \frac{1}{\sqrt{2 N_{incoh_v}}} \begin{bmatrix} \sum_{k=1}^{N_{incoh_v}} \left\{ \sum_{m=g_{vk}}^{g_{vk}+N_{vk}-1} \frac{\hat{A}_m}{\sqrt{N_{vk}}} \cos\left(\hat{\theta}_{e_m}\right) \right\}^2 \frac{\partial \hat{R}\left(-\Delta+\tau_{e_v}\right)^2}{\partial \tau} \\ \sum_{k=1}^{N_{incoh_v}} \left\{ \sum_{m=g_{vk}}^{g_{vk}+N_{vk}-1} \frac{\hat{A}_m}{\sqrt{N_{vk}}} \sin\left(\hat{\theta}_{e_m}\right) \right\}^2 \frac{\partial \hat{R}\left(-\Delta+\tau_{e_v}\right)^2}{\partial \tau} \\ \sum_{k=1}^{N_{incoh_v}} \left\{ \sum_{m=g_{vk}}^{g_{vk}+N_{vk}-1} \frac{\hat{A}_m}{\sqrt{N_{vk}}} \cos\left(\hat{\theta}_{e_m}\right) \right\}^2 \frac{\partial \hat{R}\left(\Delta+\tau_{e_v}\right)^2}{\partial \tau} \\ \sum_{k=1}^{N_{incoh_v}} \left\{ \sum_{m=g_{vk}}^{g_{vk}+N_{vk}-1} \frac{\hat{A}_m}{\sqrt{N_{vk}}} \sin\left(\hat{\theta}_{e_m}\right) \right\}^2 \frac{\partial \hat{R}\left(\Delta+\tau_{e_v}\right)^2}{\partial \tau} \\ \sum_{k=1}^{N_{incoh_v}} \left\{ \sum_{m=g_{vk}}^{g_{vk}+N_{vk}-1} \frac{\hat{A}_m}{\sqrt{N_{vk}}} \cos\left(\hat{\theta}_{e_m}\right) \right\}^2 \frac{\partial \hat{R}\left(\tau_{e_v}\right)^2}{\partial \tau} \\ \sum_{k=1}^{N_{incoh_v}} \left\{ \sum_{m=g_{vk}}^{g_{vk}+N_{vk}-1} \frac{\hat{A}_m}{\sqrt{N_{vk}}} \sin\left(\hat{\theta}_{e_m}\right) \right\}^2 \frac{\partial \hat{R}\left(\tau_{e_v}\right)^2}{\partial \tau} \end{bmatrix} \tag{5.32}$$

The measurement noise matrix contains the covariance between different measurements. The correlation between the measurements of the early, late, and prompt signals can be found as follows. Let $n_{I_{\delta 1}}$, $n_{Q_{\delta 1}}$, $n_{I_{\delta 2}}$, and $n_{Q_{\delta 2}}$ define the noise of the in-phase and the quad-phase of two signals, where δ_1 and δ_2 are the code delays used to generate the two signals. $|\delta_1 - \delta_2|$ is the code separation between the two signals. The noise is AWGN with unit variance, so

$$E\left[n_{I_{\delta 1}}^2\right] = E\left[n_{Q_{\delta 1}}^2\right] = 1 \tag{5.33}$$

The in-phase noise and the quad-phase noise are uncorrelated, so

$$E\left[n_{I_{\delta 1}}\, n_{Q_{\delta 1}}\right] = 0 \tag{5.34}$$

The correlation between different signals is

$$E\left[n_{I_{\delta 1}}\, n_{I_{\delta 2}}\right] = E\left[n_{Q_{\delta 1}}\, n_{Q_{\delta 2}}\right] = 1 - \frac{|\delta_1 - \delta_2|}{T_{chip}} \tag{5.35}$$

The correlations between the squares of the noise are

$$E\left[n_{I_{\delta 1}}^2\, n_{Q_{\delta 1}}^2\right] = 1 \tag{5.36}$$

$$E\left[n_{I_{\delta 1}}^4\right] = E\left[n_{Q_{\delta 1}}^4\right] = 3E\left[n_{I_{\delta 1}}^2\right]^2 = 3E\left[n_{Q_{\delta 1}}^2\right]^2 = 3 \tag{5.37}$$

$$E\left[n_{I_{\delta 1}}^2\, n_{I_{\delta 2}}^2\right] = E\left[n_{I_{\delta 1}}^2\right] E\left[n_{I_{\delta 2}}^2\right] + 2\,E\left[n_{I_{\delta 1}}\, n_{I_{\delta 2}}\right]^2 = 1 + 2\left(1 - \frac{|\delta_1 - \delta_2|}{T_{chip}}\right)^2 \tag{5.38}$$

Similarly,

$$E\left[n_{Q_{\delta 1}}^2\, n_{Q_{\delta 2}}^2\right] = 1 + 2\left(1 - \frac{|\delta_1 - \delta_2|}{T_{chip}}\right)^2 \tag{5.39}$$

The residual of the EKF is calculated as

$$Res_v = z_v - \hat{z}_v \tag{5.40}$$

In all the EKF types, a gain K_v is calculated, and then the start time of the next received code is estimated from

$$t_{start_{v+1}} = t_{start_{m+1}} = t_{start_m} + T_{d_m} + K_v\, Res_v \tag{5.41}$$

where $t_{start_{v+1}}$ is the estimated start of the first code in the next T_{code_v} i
This is the same as the estimated start of the first code in the next T_{d_m} i
t_{start_m} is the estimated start of the first code in the last T_{d_m} interval of the last T_{code_v} interval. This is because the EKF produces propagated values every T_{d_m} sec., but it produces updated estimates after each T_{code_v} sec. If the EKF does not produce updated estimates of the code-delay error, then the time of the start of the next data bit is estimated from

$$t_{start_{m+1}} = t_{start_m} + T_{d_m} \tag{5.42}$$

5.3.3 First-Order EKF Code Tracking

Two designs of a first-order EKF are developed. The measurement vector of the first EKF design has the form

$$z_v = \begin{bmatrix} I_{Ec_v} & Q_{Ec_v} & I_{Lc_v} & Q_{Lc_v} \end{bmatrix}^T \tag{5.43}$$

where the vector's four elements are obtained as in (5.23) and (5.24). The autocorrelation function is approximated by a triangle such that

$$\hat{R}(\tau) = 1 - \frac{|\tau|}{T_{chip}} \tag{5.44}$$

where T_{chip} is the length of one code chip, and $|\tau| \leq T_{chip}$. The differentiations of the autocorrelation function in (5.26) are obtained as follows:

$$\frac{\partial \hat{R}(-\Delta + \tau_{e_v})}{\partial \tau}\Big|_{\tau_{e_v} = \tau_{e_v}^-} = \frac{1}{T_{chip}} \tag{5.45}$$

$$\frac{\partial \hat{R}(\Delta + \tau_{e_v})}{\partial \tau}\Big|_{\tau_{e_v} = \tau_{e_v}^-} = -\frac{1}{T_{chip}} \tag{5.46}$$

The measurement noise matrix is

$$R_n = \begin{bmatrix} 1 & 0 & \left(1 - 2\frac{\Delta}{T_{chip}}\right) & 0 \\ 0 & 1 & 0 & \left(1 - 2\frac{\Delta}{T_{chip}}\right) \\ \left(1 - 2\frac{\Delta}{T_{chip}}\right) & 0 & 1 & 0 \\ 0 & \left(1 - 2\frac{\Delta}{T_{chip}}\right) & 0 & 1 \end{bmatrix} \tag{5.47}$$

The second design of the EKF has a measurement function of the form

$$z_v = I_{Einc_v} + Q_{Einc_v} - I_{Linc_v} - Q_{Linc_v}$$

where I_{Einc_v}, Q_{Einc_v}, I_{Linc_v}, and Q_{Linc_v} are taken from (5.28) and obtained as in (5.29). The expected measurement is

$$\hat{z}_v = 0$$

The H_{1v} function can be found from (5.32). The autocorrelation function is the same as in (5.44). The differentiation of the square of the autocorrelation function, which is needed in this EKF design, is

$$\frac{\partial \hat{R}(\tau)^2}{\partial \tau} = \left(2 - 2\frac{\tau}{T_{chip}}\right)\frac{1}{T_{chip}}$$

The measurement noise is found from direct applications of (5.33) to (5.39), so

$$R_n = 16\left(\frac{\Delta}{T_{chip}} - \frac{\Delta^2}{T_{chip}^2}\right)$$

In both of the first-order EKF designs, the EKF equations are applied as follows:

$$P_v^- = \Phi\, P_{v-1}^+\, \Phi^T + Q_{n_v} \tag{5.48}$$

$$K_v = P_v^-\, H_{1v}^T \left(H_{1v}\, P_v^-\, H_{1v}^T + R_n\right)^{-1} \tag{5.49}$$

$$P_v^+ = (I - K_v\, H_{1v})\, P_v^- \tag{5.50}$$

where Q_{n_v} is the process noise.

5.3.4 Second-Order EKF

The second-order EKF makes use of higher orders in the Taylor series expansion. The state x and the covariance P are updated based on the following [1]:

$$\hat{x}_v^+ = \frac{E\left[x_v\, f_{x_v}(z_v|x_v)\right]}{E\left[f_{z_v}(z_v|x_v)\right]} \tag{5.51}$$

$$\hat{P}_v^+ = \frac{E\left[x_v\, x_v^T\, f_{x_v}(z_v|x_v)\right]}{E\left[f_{x_v}(z_v|x_v)\right]} - \hat{x}_v^+\, \hat{x}_v^{+\,T} \tag{5.52}$$

where f_x and f_z are the probability density functions of the state and the measurement, respectively. These equations are solved by expressing them as a power series and getting an expression with $E[h_v - \hat{h}_v]^2$. Then, they are approximated by a Taylor series. A truncated second-order filter (TS) ignores all of the central moments above the second-order. A Gaussian second-order filter (GS) accounts for the fourth central moment, assuming the density is Gaussian. These filters use the Hessian matrix of h_v within their equations. The TS filter equations are as follows:

$$P_v^- = \Phi\, P_{v-1}^+\, \Phi^T + Q_{n_v} \tag{5.53}$$

A vector b_v of size equal to the number of measurements is calculated, where the jth element is

$$b_{vj} = \frac{1}{2}\,\text{trace}\left(\frac{\partial^2 h_{v_j}(x_v^-)}{\partial x^2}\, P_v^-\right) \tag{5.54}$$

and

$$K_v = P_v^-\, H_{1v}^T\, (H_{1v}\, P_v^-\, H_{1v}^T - b_v\, b_v^T + R_n)^{-1} \tag{5.55}$$

$$P_v^+ = (I - K_v\, H_{1v})\, P_v^- \tag{5.56}$$

The GS filter equations are

$$P_v^- = \Phi\, P_{v-1}^+\, \Phi^T + Q_{n_v} \tag{5.57}$$

$$b_{vj} = \frac{1}{2}\,\text{trace}\left(\frac{\partial^2 h_{vj}(x_v^-)}{\partial x^2}\, P_v^-\right) \tag{5.58}$$

and

$$K_v = P_v^-\, H_{1v}^T\, (H_{1v}\, P_v^-\, H_{1v}^T + B_v + R_n)^{-1} \tag{5.59}$$

$$P_v^+ = (I - K_v\, H_{1v})\, P_v^- \tag{5.60}$$

where B_v is a matrix of size $J \times J$. J is the number of measurements. The element jl of B_v is

$$B_{vjl} = \frac{1}{2}\,\text{trace}\left(\frac{\partial^2 h_{vj}(x_v^-)}{\partial x^2}\, P_v^-\, \frac{\partial^2 h_{vl}(x_v^-)}{\partial x^2}\, P_v^-\right) \tag{5.61}$$

In both of the second-order EKFs, the state update equation is

$$x_v^+ = x_v^- + K_v\left(z_v - \hat{z}_v - b_v\right) \tag{5.62}$$

5.3.5 Second-Order EKF Code Tracking

Two different second-order EKF designs are developed. Both designs can apply either the TS or the GS filters described in Section 5.3.4. A second-order EKF needs the second derivative of the autocorrelation function. Therefore, the autocorrelation function is approximated by a quadratic function of the form

$$\hat{R}(\tau) = a\left(\frac{\tau}{T_{chip}}\right)^2 + b\left(\frac{\tau}{T_{chip}}\right) + c \tag{5.63}$$

The derivatives of the approximated autocorrelation function, which are needed for H_{1v} and H_{2v}, are

$$\frac{\partial\,\hat{R}(\tau)^2}{\partial\tau} = 4a^2\frac{\tau^3}{T_{chip}^4} + 6ab\frac{\tau^2}{T_{chip}^3} + 4ac\frac{\tau}{T_{chip}^2} + 2b\frac{\tau}{T_{chip}^2} + 2bc\frac{1}{T_{chip}} \tag{5.64}$$

$$\frac{\partial^2\,\hat{R}(\tau)^2}{\partial\tau^2} = 12a^2\frac{\tau^2}{T_{chip}^4} + 12ab\frac{\tau}{T_{chip}^3} + 4ac\frac{1}{T_{chip}^2} + 2b\frac{1}{T_{chip}^2} \tag{5.65}$$

The constants a, b, and c are determined by setting $\hat{R}(\tau)$ to 1 at $\tau = 0$ and to $1 - (|\Delta|/T_{chip})$ at $\tau = \Delta$ and $\tau = -\Delta$. Therefore, the constants are obtained by solving

$$\begin{bmatrix} 1 - \frac{|\Delta|}{T_{chip}} \\ 1 - \frac{|\Delta|}{T_{chip}} \\ 1 \end{bmatrix} = \begin{bmatrix} \left(-\frac{\Delta}{T_{chip}}\right)^2 & -\frac{\Delta}{T_{chip}} & 1 \\ \left(\frac{\Delta}{T_{chip}}\right)^2 & \frac{\Delta}{T_{chip}} & 1 \\ 0 & 0 & 1 \end{bmatrix} \begin{bmatrix} a \\ b \\ c \end{bmatrix} \tag{5.66}$$

The measurement vector of the first EKF design is found from (5.28) and (5.29) as

$$z_v = \frac{1}{\sqrt{2}}\begin{bmatrix} I_{Einc_v} + Q_{Einc_v} \\ I_{Linc_v} + Q_{Linc_v} \\ I_{Pinc_v} + Q_{Pinc_v} \end{bmatrix} \tag{5.67}$$

$\hat{z}_v$, H_{1v}, and H_{2v} are found from (5.31), (5.32), and (5.27), respectively. The measurement noise matrix is

$$R_n = \begin{bmatrix} 1 & \left(1 - 2\frac{\Delta}{T_{chip}}\right)^2 & \left(1 - \frac{\Delta}{T_{chip}}\right)^2 \\ \left(1 - 2\frac{\Delta}{T_{chip}}\right)^2 & 1 & \left(1 - \frac{\Delta}{T_{chip}}\right)^2 \\ \left(1 - \frac{\Delta}{T_{chip}}\right)^2 & \left(1 - \frac{\Delta}{T_{chip}}\right)^2 & 1 \end{bmatrix} \tag{5.68}$$

The measurement vector of the second EKF design is

$$z_v = \frac{1}{\sqrt{2}} \begin{bmatrix} I_{Einc_v} - I_{Linc_v} \\ Q_{Einc_v} - Q_{Linc_v} \\ I_{Pinc_v} + Q_{Pinc_v} \end{bmatrix} \tag{5.69}$$

There is no correlation between different measurements of this EKF design. Therefore, the measurement noise matrix is

$$R_n = \begin{bmatrix} 4\left(\frac{\Delta}{T_{chip}} - \frac{\Delta^2}{T_{chip}^2}\right) & 0 & 0 \\ 0 & 4\left(\frac{\Delta}{T_{chip}} - \frac{\Delta^2}{T_{chip}^2}\right) & 0 \\ 0 & 0 & 1 \end{bmatrix} \tag{5.70}$$

In both the EKF designs, the code-delay update in (5.41) is modified as

$$t_{start_{v+1}} = t_{start_{m+1}} = t_{start_m} + T_{d_m} + K_v\left(Res_v - b_v\right) \tag{5.71}$$

5.4 Carrier Tracking

5.4.1 EKF Carrier Tracking

The EKF carrier tracking uses the prompt signal, which is generated as described in Section 5.3.1. Only coherent integration is used. T_{car_u} defines the uth integration interval length; T_{car_u} can be a multiple of one data bit interval. The EKF keeps track of the received phase, the IF carrier frequency modified by the Doppler shift (or only the Doppler shift if the received signal is converted to baseband), and the Doppler rate. The EKF model and the estimation concept are similar to those described in Section 4.7. However, the difference is that the EKF developed in Section 4.7 is implemented to track the errors of the phase, Doppler shift, and Doppler rate relative to the start of the tracking. The measurement is based on generating an error signal proportional to the current carrier phase error. This error signal is generated in Section 4.7.3 by counterrotating the in-phase and the quad-phase signals by the total estimated carrier phase error. In the EKF carrier tracking introduced in this chapter, no phase rotation is needed because the EKF produces output every one data bit interval (i.e., every T_{d_m} sec., and this output is passed to the signal generator module. Therefore, the correlated signals are generated with the current best estimates of the carrier parameters.

The dynamics model in (4.55) is modified as

$$\begin{bmatrix} \theta_{u+1} \\ \omega_{d_{u+1}} \\ \alpha_{\omega_{u+1}} \end{bmatrix} = \begin{bmatrix} 1 & T_{car_u} & \frac{T_{car_u}^2}{2} \\ 0 & 1 & T_{car_u} \\ 0 & 0 & 1 \end{bmatrix} \begin{bmatrix} \theta_u \\ \omega_{d_u} \\ \alpha_{\omega_u} \end{bmatrix} + \begin{bmatrix} \omega_{IF}\, T_{car_u} \\ 0 \\ 0 \end{bmatrix} + \begin{bmatrix} W_{\theta,\,u+1} \\ W_{\omega_d,\,u+1} \\ W_{\alpha_\omega,\,u+1} \end{bmatrix} \tag{5.72}$$

where ω_{IF}, ω_d, and α_ω are the radian IF carrier frequency, Doppler shift, and Doppler rate, respectively. The state vector is

$$x_u = \begin{bmatrix} \theta_u \\ \omega_{d_u} \\ \alpha_{\omega_u} \end{bmatrix}$$

The updated and propagated x_u are defined as x_u^+ and x_u^-, respectively.

The in-phase and quad-phase prompt signals used at the uth interval are generated as

$$I_u = \frac{1}{\sqrt{N_u}} \sum_{m=g_u}^{g_u+N_u-1} \left[\left(1 - F_{d_m}\right) \hat{d}_m + F_{d_m}\, F_{slip_u}\, d_{L_m} \right] I_m(0) \tag{5.73}$$

$$Q_u = \frac{1}{\sqrt{N_u}} \sum_{m=g_u}^{g_u+N_u-1} \left[\left(1 - F_{d_m}\right) \hat{d}_m + F_{d_m}\, F_{slip_u}\, d_{L_m} \right] Q_m(0) \tag{5.74}$$

where N_u is the number of data bits in the uth interval, g_u is the index of the first data bit in the uth interval, $\hat{d}_m$ is the estimated data bit value, which is obtained from the VA. The estimated average signal level over the uth interval, is calculated from

$$\hat{A}_u = \sqrt{I_u^2 + Q_u^2} \tag{5.75}$$

5.4.2 Square Root EKF

The EKF can suffer from instability problems due to numerical errors [2]. The covariance matrix must be symmetric and positive definite. The gain calculation equation, (5.49), requires matrix multiplication and inversion. If n is the number of measurements, then $n \times n$ matrix inversion is required. The covariance update equation, (5.50), uses the gain, K, in its calculation. If these calculations are done using finite word length, then numerical precision problems can arise. The updated covariance matrix, (5.50), is not assured to maintain its symmetric- and positive-definiteness properties. If the covariance matrix is no longer positive definite, then it will have negative eigenvalues; once this happens, all subsequent calculations will be erroneous [2]. More discussion on this subject can be found in [2, Chapters 5 and 7].

The numerical precision problem can be circumvented by adapting alternate forms for the calculations of the gain and the covariance matrix update. These alternate forms belong to a version of the EKF called the square root EKF, which exhibits improved numerical precision and stability [2]. The covariance update equation is calculated in a way that maintains its properties, as can be seen in the next section.

5.4.3 Square Root EKF Carrier Tracking

The carrier- and code-tracking modules use each other's output. Thus, the EKFs of the two modules depend on each other. If one of the EKFs loses lock, the other will too. To maximize the likelihood of maintaining lock, a stable EKF is needed. Therefore, a square root EKF for carrier tracking is implemented.

The square root EKF implements the Carlson algorithm [2]. Define S_u^- and S_u^+ as the propagated and the updated square root of the covariance. S_u^- is generated as follows:

$$X_u = \Phi \, S_{u-1}^+ \tag{5.76}$$

$$P_u^- = X_u \, X_u^T + Q_{n_u} \tag{5.77}$$

$$S_u^- = chol(P_u^-) \tag{5.78}$$

where $chol(P^-)$ is the Cholesky square root matrix of the propagated covariance, and Φ and Q_n are, respectively, the transition and the process noise matrices calculated as in Chapter 4, Section 4.7.2. The Carlson algorithm finds S_u^+ by solving the following equation [2]:

$$S_u^+ = S_u^- \left[\mathbf{I} - \frac{a_u \, a_u^T}{\beta_u} \right] \tag{5.79}$$

where **I** is a unit matrix, and

$$a_u = S_u^{-^T} H_u^T \tag{5.80}$$

$$\beta_u = a_u \, a_u^T + R_n \tag{5.81}$$

The Carlson algorithm has the advantage of keeping the triangularity of S_u^+, which is calculated column-by-column iteratively. Define n as the number of states (i.e., phase, Doppler, and rate). The algorithm is initialized as follows:

$$\beta_0 = R_n \tag{5.82}$$

$$e_0 = 0 \tag{5.83}$$

$$a = S_u^{-T} H_u^T \tag{5.84}$$

Then, the following are calculated for $k = 1, \ldots, n$:

$$\beta_k = \beta_{k-1} + a_k^2 \tag{5.85}$$

$$u_k = (\beta_{k-1}/\beta_k)^{1/2} \tag{5.86}$$

$$v_k = a_k/(\beta_{k-1}\,\beta_k)^{1/2} \tag{5.87}$$

$$e_k = e_{k-1} + S_k^- a_k \tag{5.88}$$

$$S_k^+ = S_k^- u_k - e_{k-1} v_k \tag{5.89}$$

where a_k is the kth element of the vector a defined in (5.84), e_k is a vector, S_k^+ is a column vector, and S_u^+ is found as

$$S_u^+ = [S_1^+ \ \ S_2^+ \ \ \cdots \ \ S_n^+] \tag{5.90}$$

The gain is calculated from

$$K_u = \frac{e_n}{\beta_n} \tag{5.91}$$

The measurement is taken as the quad-phase signal Q_u. So,

$$z_u = Q_u \equiv A_u \sin(\theta_{e_u}) \tag{5.92}$$

where A_u is the average signal level over the uth interval. θ_{e_u} is the average carrier phase error. The expected measurement is

$$\hat{z}_u = 0 \tag{5.93}$$

The first derivative of h_u, where $z_u = h_u + n_u$, produces the following H_u function:

$$H_u = [\hat{A}_u \ \cos(\theta_{e_u}^-) \ \ 0 \ \ 0] = [\hat{A}_u \ \ 0 \ \ 0] \tag{5.94}$$

where $\hat{A}$ is calculated as in (5.75). The residual is

$$Res_u = Q_u \tag{5.95}$$

5.4.4 Second-Order EKF Carrier Tracking

A second-order EKF is designed. The EKF can implement either a TS or a GS filter as previously described in Section 5.3.4. The measurement is

$$z_u = \begin{bmatrix} Q_u \\ I_u \end{bmatrix} \equiv \begin{bmatrix} A_u \ \sin(\theta_{e_u}) \\ A_u \ \cos(\theta_{e_u}) \end{bmatrix} \tag{5.96}$$

where Q_u and I_u are calculated as in (5.73) and (5.74). The expected measurement is

$$\hat{z}_u = \begin{bmatrix} 0 \\ \hat{A}_u \end{bmatrix} \tag{5.97}$$

The first derivative of the h_u function produces the following H_{1u} function:

$$H_{1u} = \begin{bmatrix} \hat{A}_u & 0 & 0 \\ 0 & 0 & 0 \end{bmatrix} \tag{5.98}$$

The second derivative of the h_u function produces two matrices:

$$H_{21u} = \frac{\partial^2 h_{u1}(x_u^-)}{\partial^2 x} = \begin{bmatrix} 0 & 0 & 0 \\ 0 & 0 & 0 \\ 0 & 0 & 0 \end{bmatrix} \tag{5.99}$$

$$H_{22u} = \frac{\partial^2 h_{u2}(x_u^-)}{\partial^2 x} = \begin{bmatrix} -\hat{A}_u & 0 & 0 \\ 0 & 0 & 0 \\ 0 & 0 & 0 \end{bmatrix} \tag{5.100}$$

where h_{u1} and h_{u2} are the first and the second rows of h_u, respectively.
The EKF equations are applied as in Section 5.3.4. The state update equation is

$$x_u^+ = x_u^- + K_u(z_u - \hat{z}_u - b_u) \tag{5.101}$$

where b_u is determined from (5.54).

5.4.5 Second-Order Square Root EKF Carrier Tracking

The Potter algorithm, described in [2], works with a first-order filter. It is modified in this section to work with a second-order filter. The Potter algorithm differs from the Carlson algorithm in the covariance update equations, but it uses (5.76) to (5.81). The covariance update for a first-order EKF is found as follows [2]:

$$\alpha_u = \frac{1}{a_u^T\, a_u + R_n} \tag{5.102}$$

$$\gamma_u = \frac{1}{1 + (\alpha_u \;\; R_n)^{1/2}} \tag{5.103}$$

$$K_u = \alpha_u\, S_u^-\, a_u \tag{5.104}$$

$$S_u^+ = S_u^- - \gamma_u\, K_u\, a_u^T \tag{5.105}$$

The square root second-order EKF uses two independent measurements as in Section 5.4.4. Thus, (5.102) to (5.105) are modified as follows:

$$\alpha_{ku} = \frac{1}{a_{ku}^T\, a_{ku} + R_{n_k} + B_{ku}} \tag{5.106}$$

$$\gamma_{ku} = \frac{1}{1 + (\alpha_{ku} \;\; (R_{n_k} + B_{ku}))^{1/2}} \tag{5.107}$$

$$K_{ku} = \alpha_{ku}\, S_u^-\, a_{ku} \tag{5.108}$$

$$K_u = [K_{1u}\; K_{2u}] \tag{5.109}$$

$$\kappa_u = [\gamma_{1u} K_{1u}\, \gamma_{2u}\, K_{2u}] \tag{5.110}$$

$$S_u^+ = S_u^- - \kappa_u\, a_u^T \tag{5.111}$$

where a_{ku} is the kth column of a_u, R_{n_k} is the kth diagonal element of the measurement noise matrix, and B_{ku} is the kth diagonal element of B_u, which is found from (5.61).

5.5 Navigation Message Decoding

The algorithms developed in this book deal with very low C/N_0 signals. Therefore, the theoretical BER is high and does not allow for the decoding of the navigation message. An algorithm is developed in Section 4.8 to detect the data bits under weak signal conditions; this algorithm utilizes the fact that some contents of the navigation message repeat after every specified interval. In this section, the navigation message structure is utilized to detect and decode the navigation message. The repeated detected bits are passed to the code- and carrier-tracking modules to be used in subsequent integrations to remove the data signs. Therefore, the coherent integration can be increased and consequently the code- and carrier-tracking performances can be enhanced.

The navigation message [3] consists of five subframes. Each subframe has a 6-second duration; it starts with a known 8-bit preamble and consists of 10 30-bit words. The subframe ID occupies bits 20 to 22 in the second word. As described in Section 1.10, each word can be decoded using the parity in bits 25 to 30 of that word and the last two bits of the previous word. This means that bit errors in other words, except for the last two bits of the previous word, will not affect the decoding of that word. Thus, each word can be decoded independently of the other words.

5.5.1 Finding the Preamble

The preamble marks the beginning of each subframe, so it repeats every 6 seconds. The following steps are applied to find the preamble:

1. Detect 6 seconds of data bits using the VA.
2. Search for all of the preamblelike sequences.
3. If no sequence is found, go back to step 1 to detect the next 6 seconds' worth of data.
4. For each found sequence, test the parity bits of the 30-bit word that starts with that sequence.
5. If the parity check test passes, then this signifies that the preamble has been found. So, the algorithm moves to detect the subframe ID.
6. If the parity check test fails for the word that starts with that sequence, then check all the previously detected 6 seconds of data with the same index as that sequence.
7. Count the number of times the preamble appears; define this number as $N_{preamble}$.
8. If $N_{preamble} < N_{min}$, where N_{min} is a predefined number, then go back to step 1.
9. If $N_{preamble} \geq N_{min}$, then check all the received data bits located at all the subframe IDs (i.e., the data bits located at the indices 50 to 52 relative to the start of the found preamble).
10. If the subframe IDs do not follow a correct pattern, then they ID cannot be concluded (this is explained in Section 5.5.2). Therefore, go back to step 1.

11. If the subframe IDs follow a correct pattern, then conclude both the preamble and the subframe IDs (i.e., the start of each subframe as well as their IDs are correctly defined).

5.5.2 Identifying the Subframe IDs

The subframe IDs are identified by either of two methods: (1) decoding the second word of a subframe, which can be done after finding the preamble; (2) searching for a correct ID pattern, which can be done either in conjunction with the search for the preamble or after finding it. Each subframe has a distinct 3-bit ID. Each ID refers to its subframe index within the five subframes of the navigation message. This means that there are five possibilities for the order of the IDs in a sequence of five consecutive subframes. Each possibility assumes that a different subframe is the first one in the sequence. Thus, there can only be five patterns of IDs in any consecutive five subframes. These patterns are

$$pattern_1 = [001\ 010\ 011\ 100\ 101]$$

$$pattern_2 = [010\ 011\ 100\ 101\ 001]$$

$$pattern_3 = [011\ 100\ 101\ 001\ 010]$$

$$pattern_4 = [100\ 101\ 001\ 010\ 011]$$

$$pattern_5 = [101\ 001\ 010\ 011\ 100]$$

where, $pattern_i$ assumes that the first subframe in the sequence is i. The first method to identify the IDs works as follows:

1. If the preamble has just been identified, then try to decode the second word of all the received subframes.
2. If a word can be correctly decoded, then the ID of the subframe of which this word is a part is identified as bits 20 to 22 of that word. If the data are decoded with the correct signs, then bits 20 to 22 contain the correct ID signs. Otherwise, this subframe ID is identified as of the reverse sign of bits 20 to 22. The other four subframes' IDs can be directly constructed based on the identified subframe ID.
3. If no word can be correctly decoded, then go to the second method to search for a correct ID pattern.
4. If both methods fail to identify the IDs, then wait for the reception of the second word of the next subframe and try to decode it. Then go back to step 2.

The second method to identify the IDs is to search for a correct pattern of IDs. Let N_{sub} define the number of subframes. n_{sub} defines a subframe index in the range of 1 to 5; this index refers to the order of a subframe within the received sequence of subframes. n_{ID} defines the real subframe number. Five matrices, $M_{pattern_i}$, are generated. Each matrix has a size of $N_{sub} \times N_{sub}$. Each matrix corresponds to a possible order of the received subframes (i.e., matrix i assumes that the first received subframe number is i). Each row represents the index of a subframe within a received sequence, while each column represents the real subframe number. Each matrix contains 1's at the indices that represent the intersection of a subframe index within a received sequence and its real number (n_{sub}, n_{ID}); it contains zeros otherwise. These matrices are

$$M_{pattern_1} = \begin{bmatrix} 1 & 0 & 0 & 0 & 0 \\ 0 & 1 & 0 & 0 & 0 \\ 0 & 0 & 1 & 0 & 0 \\ 0 & 0 & 0 & 1 & 0 \\ 0 & 0 & 0 & 0 & 1 \end{bmatrix} \tag{5.112}$$

$$M_{pattern_2} = \begin{bmatrix} 0 & 1 & 0 & 0 & 0 \\ 0 & 0 & 1 & 0 & 0 \\ 0 & 0 & 0 & 1 & 0 \\ 0 & 0 & 0 & 0 & 1 \\ 1 & 0 & 0 & 0 & 0 \end{bmatrix} \tag{5.113}$$

$$M_{pattern_3} = \begin{bmatrix} 0 & 0 & 1 & 0 & 0 \\ 0 & 0 & 0 & 1 & 0 \\ 0 & 0 & 0 & 0 & 1 \\ 1 & 0 & 0 & 0 & 0 \\ 0 & 1 & 0 & 0 & 0 \end{bmatrix} \tag{5.114}$$

$$M_{pattern_4} = \begin{bmatrix} 0 & 0 & 0 & 1 & 0 \\ 0 & 0 & 0 & 0 & 1 \\ 1 & 0 & 0 & 0 & 0 \\ 0 & 1 & 0 & 0 & 0 \\ 0 & 0 & 1 & 0 & 0 \end{bmatrix} \tag{5.115}$$

$$M_{pattern_5} = \begin{bmatrix} 0 & 0 & 0 & 0 & 1 \\ 1 & 0 & 0 & 0 & 0 \\ 0 & 1 & 0 & 0 & 0 \\ 0 & 0 & 1 & 0 & 0 \\ 0 & 0 & 0 & 1 & 0 \end{bmatrix} \tag{5.116}$$

A matrix, *orderID*, of size $N_{sub} \times N_{sub}$ is generated; the ith row contains the order of the received subframes that is expressed in $M_{pattern_i}$. This matrix is

$$orderID = \begin{bmatrix} 1 & 2 & 3 & 4 & 5 \\ 2 & 3 & 4 & 5 & 1 \\ 3 & 4 & 5 & 1 & 2 \\ 4 & 5 & 1 & 2 & 3 \\ 5 & 1 & 2 & 3 & 4 \end{bmatrix} \tag{5.117}$$

The second method works as follows:

1. Extract the preamble and bits 20 to 22 of the second word from all the received subframes.
2. If the preamble's subframe BER is less than or equal to a predefined threshold $\gamma_{preamble}$, then store both the ID and its index, n_{sub}.
3. Generate a matrix M_{sub} of size $N_{sub} \times N_{sub}$, and set it to zero.
4. For each of the stored IDs, check that they match any of the real IDs. If a match occurs, then store 1 in M_{sub} at the index (n_{sub}, n_{ID}), where n_{ID} refers to the matched ID number.
5. Multiply M_{sub} by each of the $M_{pattern_i}$ matrices elementwise. Add all the elements of each resultant matrix. Store the addition result in a vector Λ, where the ith result is stored in the ith element of Λ, defined as Λ_i.
6. Define the maximum element of Λ as Λ_m and its index as m. If Λ_m exceeds a predefined threshold ID_{th}, given that no other elements in Λ exceed it, then the subframe identification is concluded. The received subframe ID order is expressed in the mth row of the matrix *orderID*.
7. If the subframe ID cannot be concluded, then wait for the reception of the second word of the next subframe, and then go back to method 1.

5.5.3 Decoding Repeated Subframe Data

Words 3 to 10 of subframes 1 to 3 repeat every 30 seconds, while words 3 to 10 of subframes 4 and 5 repeat every 12.5 minutes. A method is developed to detect the repeated data bits of very weak signals, which otherwise might never be detected.

After the subframe IDs have been defined, the following variables are generated: a counter *NavData*, which keeps track of the current data bit index within its subframe; a matrix M_{word} of size $N_{sub} \times N_{words}$, which keeps track of the repeated words that have been correctly detected, where N_{words} is the number of words in a subframe and in which 1 is stored at the indices that represent correctly decoded words; a matrix M_{tSig} of size $N_{sub} \times N_{bits}$, which stores the summation of the signed signal levels of the repeated bits that have not yet been detected, where N_{bits} is the number of data bits in a subframe; a matrix M_{bit} of size $N_{sub} \times N_{bits}$, which stores the correctly decoded bits, d_{L_m}; and a vector V_{sig} of size N_{bits}, which stores the signed signal levels of the current received subframe (i.e., $\hat{d}_m \hat{A}_m$), which is taken from the output of the carrier-tracking module. All these variables, except for M_{bit} and V_{sig}, are kept until all the repeated data are detected, and then they are deleted.

The general idea of this method is to add the signed signal levels of the repeated data that have not yet been detected. The addition of the signed signal levels increases the SNR, as shown in Section 4.8. The increase in the SNR causes a decrease in the BER for a given C/N_0. However, the problem is that there can be undetected 180° carrier phase error in some of the repeated data. Such a problem can cause the total signed signal level to go to zero, which will prevent the correct detection of the data. This problem is circumvented as follows. The total summation of the signed signal levels of the previously received repeated data is stored in M_{tSig}. The signed signal levels of the currently received repeated subframe are stored in V_{sig}. At the end of the reception of a complete repeated word, the signs of the signal levels of that word in both M_{tSig} and V_{sig} are compared, and the number of the sign disagreements is counted; define this number as $N_{signDisAg}$. If $N_{signDisAg}/30$ is less than a threshold γ_{level}, then the signed signal levels of that word in V_{sig} are added to the total signed signal levels of the corresponding word in M_{tSig}; otherwise, the signs of the signal levels in V_{sig} are reversed before they are added to the signed signal levels in M_{tSig}. The addition results are stored in M_{tSig}.

This method will enable the detection of the data bits with correct or reversed signs, depending on the signs of the data in the first received word. However, the detected data will be correctly decoded whether the detection is done with correct or reversed signs [3]. The method works as follows:

1. Store the signed signal levels of the current received word in V_{sig}.
2. At the end of the reception of a complete word, if that word has not been decoded, then try to decode it using the decoding algorithm in [3]. If it can be correctly decoded, then store its detected data in the corresponding locations in M_{bit}.
3. If the word cannot be decoded, then add the signed signal levels of that word in V_{sig} to those in M_{tSig}. Then, try to decode that word from the signed signal levels in M_{tSig}.
4. If the word is correctly decoded, then its data bits are stored in M_{bit}.
5. If the current received word has previously been decoded, then F_{d_m} of the current data bit is set to 1, and d_{L_m} is extracted from M_{bit}. Thus, the correct data signs are used within EKFs of the code and carrier tracking.

If there is a 180° carrier phase error, then F_{slip} is set to -1; otherwise F_{slip} is set to 1. This is to avoid using data with incorrect signs relative to each other if the integration interval contains some data bits that have already been detected and other data bits that have not been detected. The signs of the current detected data bits, $\hat{d}_m$, are extracted from V_{sig}; these signs are compared to the signs of the correct data, which are stored in M_{bit} to determine whether there is a 180° carrier phase error or not. The EKF tracking can be degraded if the current data bits are being detected with reversed signs and the correct signs are used.

5.6 Maximizing the Time to Lose Lock

One of the main factors affecting the time to lose lock is the noise variance of the tracking modules. Increasing the integration time reduces the noise variance. The coherent integration has better sensitivity as compared to the incoherent integration, but it has some limitations. A frequency error of f_e Hz introduces a loss proportional to sinc(T_I f_e) when a coherent integration length of T_I sec. is used. The BER of the estimated data introduces another loss if the coherent integration is larger than one data bit interval. In this section, some techniques are implemented to reduce the noise variance and increase the time that the receiver remains in track mode.

In addition, the tracking performance is monitored to detect large Doppler shift or code-delay errors. A large Doppler shift error increases the BER and degrades the signal level. A large code-delay error degrades the signal level. If a large Doppler shift error is concluded, then a carrier reinitialization module is activated to refine the error. If a large code-delay error is concluded, then an acquisitionlike module is activated. The operations of these two modules are explained in Sections 5.6.3 and 5.6.4.

5.6.1 Adaptive Integration Time

5.6.1.1 Code Tracking

The code-tracking module is implemented to have the coherent, incoherent, and the total integration lengths change with time, based on the tracking conditions. The coherent integration is set to a maximum value if the current data have been correctly detected and the frequency-tracking error is small. In contrast, it is set to a minimum value if the current data have not been correctly detected, or if there is a large frequency error. The total integration, on the other hand, is set to a small value to achieve rapid convergence in the EKF, and it is set to a large value to reduce the noise variance.

Four integration lengths are set prior to the start of the tracking. Since the data length, T_{d_m}, is not constant over each interval, the integration lengths are defined as multiples of one data bit. These lengths are the minimum total integration, the maximum total integration, the minimum coherent integration, and the maximum coherent integration; they are defined as T_{tmin}, T_{tmax}, $T_{I_{Min}}$, and $T_{I_{Max}}$, respectively, and they are multiple N_{tmin}, N_{tmax}, $N_{I_{Min}}$, and $N_{I_{Max}}$ of one data bit interval, respectively.

Define N_{code_v} as the number of the data bit intervals in the vth interval, and N_{Icod_v} as the number of data bit intervals in the coherent integration at the vth interval. The adaptive integration works as follows:

1. Start the code-tracking module by setting $N_{code_v} = N_{tmin}$ and $N_{Icod_v} = N_{I_{Min}}$.
2. Gradually increase the total integration length with time. This is done by defining two vectors. The first contains the amount of time between the increase in the integration length, and the second contains the amount of the increase.
3. Before calculating z, $\hat{z}$, and H, the coherent and incoherent integrations are adjusted. T_{code_v} consists of N_{incoh_v} coherent integrations that are added incoherently. The length of each of the N_{inoch_v} coherent integrations does not have to be the same, but there can be up to three different coherent integration lengths, and each can be a multiple of one data bit interval. These lengths are as follows: (a) T_{d_m} for the intervals whose data have not yet been detected (i.e., the intervals that have $F_{d_m} = 0$); (b) Multiple $\min\{N_{I_{Max}}, N_{dc_v}\}$ of one data bit interval, where N_{dc_v} is the number of correctly decoded data in the current T_{code_v} sec.; and (c) the remaining T_{d_m} intervals with correctly decoded data that have total length less than multiple $\min\{N_{I_{Max}}, N_{dc_v}\}$ of one data bit interval (i.e., multiple $[\min\{N_{I_{Max}}, N_{dc_v}\} \bmod N_{code_v}]$ of one data bit interval).
4. If no data bits have been detected or the frequency error is large, then the coherent integration length is set to a multiple $N_{I_{Min}}$ of one data bit interval.

5.6.1.2 Carrier Tracking

The carrier tracking uses coherent integration only. The integration time is implemented to change adaptively, depending on the correctly decoded data. This is done as follows. The minimum and the maximum integration lengths are set as $T_{I_{cMin}}$ and $T_{I_{cMax}}$, respectively. Each integration length can be a multiple of one data bit interval, so they are multiples $N_{I_{cMin}}$ and $N_{I_{cMax}}$ of one data bit interval, respectively.

Define N_{car_u} as the number of data bit intervals in the uth interval. At the beginning of the tracking, T_{car_u} is set as a multiple $N_{I_c Min}$ of one data bit interval (i.e., $N_{car_u} = N_{I_c Min}$). After the end of each carrier integration interval, $T_{car_{u+1}}$ is set, based on the number of correctly detected data bits, if any, that start at the next interval, on the condition that

$$N_{I_c Min} \leq N_{car_{u+1}} \leq N_{I_{cMax}} \tag{5.118}$$

If the new integration length, $T_{car_{u+1}}$, is different from the old integration length, T_{car_u}, then Φ and Q_n matrices of the EKF are set based on $T_{car_{u+1}}$. These matrices can be calculated once for each possible integration length and stored to be used directly when needed. The covariance matrix P and the square root covariance matrix S are handled differently. For each possible integration length, a different P is updated and propagated. If a new integration length is set, then the last P^+, based on the old T_{car_u}, is stored to be used the next time this integration length is used, while the stored P^+ for the new $T_{car_{u+1}}$ length is used. If an integration length is being used for the first time, then P^+ of the old integration is used.

Also, if a new integration length is set, then the propagated x^- values are recalculated from the updated x^+ values to refer to those values at the middle of the next $T_{car_{u+1}}$ sec. This is done as follows:

$$\begin{bmatrix} \theta_{u+1}^- \\ \omega_{d_{u+1}}^- \\ \alpha_{\omega_{u+1}}^- \end{bmatrix} = \begin{bmatrix} 1 & \frac{T_{car_{u+1}}}{2} & \frac{T_{car_{u+1}}^2}{6} \\ 0 & 1 & \frac{T_{car_{u+1}}}{2} \\ 0 & 0 & 1 \end{bmatrix} \begin{bmatrix} \theta_{s,u+1} \\ \omega_{d_{s,u+1}} \\ \alpha_{\omega_{s,u+1}} \end{bmatrix} + \begin{bmatrix} \frac{\omega_{IF} T_{car_{u+1}}}{2} \\ 0 \\ 0 \end{bmatrix} + \begin{bmatrix} W_{\theta,u+1} \\ W_{\omega_d,u+1} \\ W_{\alpha_\omega,u+1} \end{bmatrix} \tag{5.119}$$

where $\theta_{s,u+1}$, $\omega_{d_{s,u+1}}$, and $\alpha_{\omega_{s,u+1}}$ are the phase, Doppler shift, and Doppler rate at the start of the $T_{car_{u+1}}$ interval. These values are obtained as

$$\alpha_{\omega_{s,u+1}} = \alpha_{\omega_u}^+ \tag{5.120}$$

$$\omega_{d_{s,u+1}} = \omega_{d_u}^+ + \alpha_{\omega_{s,u+1}} \frac{T_{car_u}}{2} \tag{5.121}$$

$$\theta_{s,u+1} = \theta_u^+ + \left(\omega_{d_{s,u+1}} + \omega_{IF}\right) \frac{T_{car_u}}{2} + \alpha_{\omega_{s,u+1}} \frac{T_{car_u}^2}{6} \tag{5.122}$$

5.6.2 Adaptive Early/Late Code-Delay Separation

A large code-delay separation between the early and late signals can help to maintain a code lock if the expected code-delay estimation error is large. In contrast, a small code-delay separation can give a better code-delay estimation. A method is developed to take advantage of various separations. The method enables the use of more than one code-delay separation at the same time to generate the early and late correlated signals. It also has the ability to change the delay separation adaptively.

Two approaches are introduced; the first uses a separate EKF for each code-delay separation, while the second uses only one EKF for all of the code-delay separations.

Let N_{Δ_v} define the number of the code-delay separations that are used at the vth integration interval. The first approach uses N_{Δ_v} different EKFs, one for each code-delay separation. Define Δ_j as the jth code-delay separation; Δ_j is equal to half the code-delay separation between the early and late signals. Each EKF generates independent residual and H functions, but only one update/propagate step is performed to estimate the current code-delay error, τ_{e_v}, and the start of the next integration interval. This is done by calculating an average of the $(K_v\ Res_v)$ values. Thus, (5.41) is modified to

$$t_{start_{v+1}} = t_{start_{m+1}} = t_{start_m} + T_{d_m} + \frac{1}{N_{\Delta_v}} \sum_{j=1}^{N_{\Delta_v}} K_{v_j}\ Res_{v_j} \tag{5.123}$$

where K_{v_j} and Res_{v_j} are, respectively, the gain and residual of the jth EKF at the vth interval.

The second approach uses one EKF for all of the code-delay separations. In this case, z_v, $\hat{z}_v$, H_{1v}, H_{2v}, and R_n are extended to include measurements from all of the code-delay separations. For example, if two delay separations, Δ_1 and Δ_2, are used with a second-order EKF, then (5.67) becomes

$$z_v = \frac{1}{\sqrt{2}} \begin{bmatrix} I_{Einc_v}(-\Delta_1) + Q_{Einc_v}(-\Delta_1) \\ I_{Linc_v}(\Delta_1) + Q_{Linc_v}(\Delta_1) \\ I_{Pinc_v} + Q_{Pinc_v} \\ I_{Einc_v}(-\Delta_2) + Q_{Einc_v}(-\Delta_2) \\ I_{Linc_v}(\Delta_2) + Q_{Linc_v}(\Delta_2) \\ I_{Pinc_v} + Q_{Pinc_v} \end{bmatrix} \tag{5.124}$$

Two measurements from the same prompt signal are used because of the following. The second-order EKF approximates the autocorrelation function by (5.63). For each code-delay separation, (5.66) is solved to produce different values for the constants (a, b, c). Since these constants are used in $\hat{z}_v$, H_{1v}, and H_{2v}, the residual and the gain will have different values at each of the two prompt measurements. Define

$$\Delta_{diff} = |\Delta_1 - \Delta_2| \tag{5.125}$$

$$\Delta_{sum} = \Delta_1 + \Delta_2 \tag{5.126}$$

The measurement noise matrix has a size of 6 × 6. It is as follows:

$$R_n = \begin{bmatrix} 1 & \left(1-\frac{2\Delta_1}{T_{chip}}\right)^2 & \left(1-\frac{\Delta_1}{T_{chip}}\right)^2 & \left(1-\frac{\Delta_{diff}}{T_{chip}}\right)^2 & \left(1-\frac{\Delta_{sum}}{T_{chip}}\right)^2 & \left(1-\frac{\Delta_1}{T_{chip}}\right)^2 \\ \left(1-\frac{2\Delta_1}{T_{chip}}\right)^2 & 1 & \left(1-\frac{\Delta_1}{T_{chip}}\right)^2 & \left(1-\frac{\Delta_{sum}}{T_{chip}}\right)^2 & \left(1-\frac{\Delta_{diff}}{T_{chip}}\right)^2 & \left(1-\frac{\Delta_1}{T_{chip}}\right)^2 \\ \left(1-\frac{\Delta_1}{T_{chip}}\right)^2 & \left(1-\frac{\Delta_1}{T_{chip}}\right)^2 & 1 & \left(1-\frac{\Delta_2}{T_{chip}}\right)^2 & \left(1-\frac{\Delta_2}{T_{chip}}\right)^2 & 1 \\ \left(1-\frac{\Delta_{diff}}{T_{chip}}\right)^2 & \left(1-\frac{\Delta_{sum}}{T_{chip}}\right)^2 & \left(1-\frac{\Delta_2}{T_{chip}}\right)^2 & 1 & \left(1-\frac{2\Delta_2}{T_{chip}}\right)^2 & \left(1-\frac{\Delta_2}{T_{chip}}\right)^2 \\ \left(1-\frac{\Delta_{sum}}{T_{chip}}\right)^2 & \left(1-\frac{\Delta_{diff}}{T_{chip}}\right)^2 & \left(1-\frac{\Delta_2}{T_{chip}}\right)^2 & \left(1-\frac{2\Delta_2}{T_{chip}}\right)^2 & 1 & \left(1-\frac{\Delta_2}{T_{chip}}\right)^2 \\ \left(1-\frac{\Delta_1}{T_{chip}}\right)^2 & \left(1-\frac{\Delta_1}{T_{chip}}\right)^2 & 1 & \left(1-\frac{\Delta_2}{T_{chip}}\right)^2 & \left(1-\frac{\Delta_2}{T_{chip}}\right)^2 & 1 \end{bmatrix} \quad (5.127)$$

The code-delay separation can be set to a large value at the beginning of the tracking and immediately after a reacquisition process. Otherwise, either it can be set to a small value or more than one code-delay separation can be used.

5.6.3 Reinitialization of the Carrier Parameters

The EKF of the carrier tracking will converge if the Doppler shift estimation error is less than 12 Hz. However, it might take a long time for the convergence to occur if the error is larger than about 3 to 6 Hz. During that time, the data bit estimation will degrade. Since the EKF of the code tracking uses the output of the carrier tracking, such an error will have some impact on the code-tracking performance. If the Doppler shift estimation error starts to go beyond 12 Hz, then the EKF for the carrier tracking might diverge, which will lead to a divergence in the code tracking as well. To minimize such a possibility, a carrier reinitialization module is implemented.

A large Doppler shift error is concluded if the estimated BER, $\hat{\beta}$, is high over a large number of words that is, $\hat{\beta} > \zeta\ \bar{\beta}$, given that $\bar{\beta}$ is available and $\zeta > 1$. $\bar{\beta}$ is calculated as discussed in Section 5.2. Although both the code- and the carrier-tracking modules use the correctly decoded repeated data to remove the effect of the data signs, the modules still keep track of the current estimated signed signal levels (i.e., $\hat{d}_m\ \hat{A}_m$) in the vector V_{sig}. The BER can be calculated by comparing the stored data bits in M_{bit} to the signs of the data in V_{sig}. Once a large error is concluded, the EKF of the carrier tracking is deactivated, and the ES-VA for fine acquisition, described in Sections 4.5 and 4.6, is activated. Following that, the estimated phase, Doppler shift, and Doppler rate are adjusted, and then the EKF of the carrier tracking is reactivated.

This process is controlled by continuously checking $\hat{\beta}$ of the correctly decoded words. Each time $\hat{\beta}$ is larger than $\zeta\ \bar{\beta}$, a counter, C_{BER}, is increased, but this counter is set to 0 if $\hat{\beta}$ is less than $\zeta\ \bar{\beta}$. When the counter reaches a predefined value of N_{BER}, a high Doppler shift error is concluded.

5.6.4 Reacquisition of the Code Delay

An acquisitionlike module is implemented within the code-tracking module. Its purpose is to reduce the code-delay error if it becomes large and, consequently, difficult for the EKF to converge. Since the carrier- and the code-tracking EKFs use each other's output, then a high estimation error in either of the EKFs will lead to a high error in the other. A high Doppler-shift-estimation error can be concluded as described in Section 5.6.3. Following a carrier-tracking reinitialization, a test is performed to determine whether a code reacquisition is necessary. Since a large code-delay error causes a reduction in the received signal level, the estimated signal levels over $N_{codeAcq}$ data bit intervals are compared to a previously estimated signal level. If there are large reductions in the signal levels, then the acquisitionlike process is activated.

The acquisitionlike process works over a small number of code delays around the current delay estimate. During this process, the EKF of the code tracking is deactivated. All of the code delays used in this process will correlate with the local code with different values depending on their separation from the correct code delay. An approach is used to estimate the correct code delay with a resolution less than half the sampling time and to adjust the total acquisition time adaptively. This approach uses the ratio between the accumulation output at different code delays either to conclude or to continue the acquisition process. Following the acquisition process, the EKF of the code tracking is reactivated, and the integration time and the early/late separation are adjusted, if necessary.

5.7 Handling of Random Sudden Changes in the Doppler Shift or the Doppler Rate

The Doppler shift depends on the relative velocity, v_{sr}, of the satellite and the receiver, while the Doppler rate depends on the acceleration, a_{sr}. If v_{sr} or a_{sr} changes, then a dynamic model can be included to track such a change, but if v_{sr} or a_{sr} changes in a random, sudden manner, then there are two ways to handle such a situation. The first approach is to reacquire the signal using the approaches developed in Chapter 3, Sections 3.2 and 3.3. This will be preferable in situations where those variables do not often change much, or the acquisition time is much smaller than the time between the change in those values. The other approach applies if the change occurs very often, maybe every few seconds or so. In this case the carrier-tracking algorithm is modified as follows:

1. Define the range of the possible α values and the possible f_d values as, respectively, $(\alpha_{a1}, \alpha_{a2})$ and (f_{dv1}, f_{dv2}). Let N_{va} define the number of possible combinations of f_d and α.
2. In each step of the tracking, N_{va} groups of the correlated signals (early, late, and prompt) are formed. One group, defined as G_c, assumes that no change has occurred in v_{sr} or a_{sr}, while the other $N_{va} - 1$ groups assume that a change has occurred.
3. The signal level $\hat{A}$ of each result is compared to a previously estimated signal level $\bar{A}$, where $\bar{A}$ is taken from a previous step in which both the code- and carrier-tracking modules work correctly. If the signal level $\hat{A}$ of G_c is close to $\bar{A}$, then the other groups are discarded, and the tracking is continued. Otherwise, the signal levels of the other groups are inspected to determine the group with the maximum $\hat{A}$; this group is defined as G_m.
4. Both G_c and G_m are used in the tracking for a few steps. If G_m seems to track the signal correctly, which can be concluded from the estimated signal level and the detected data bits, then G_c is replaced with G_m, and the tracking continues as in step 2. If G_m does not seem to track the signal correctly, then a large Doppler shift estimation error is concluded instead of a sudden change in v_{sr} or a_{sr}, and so the carrier reinitialization module, described in Section 5.6.3, is activated.

5.8 Computational Requirements

There are four different modules that work with signal tracking: the signal generator, code tracking, carrier tracking, and navigation message decoding. The modules generate output every T_{d_m} sec. However, the EKFs of the code and carrier tracking update their estimates every T_{code_v} sec. and T_{car_u} sec., respectively, where $T_{code_v} \geq T_{d_m}$ and $T_{car_u} \geq T_{d_m}$. The largest computations are needed in the signal generator and the EKFs of the code and carrier tracking. The computations needed by each module are as follows:

1. The signal generator generates early, prompt, and late correlated signals every T_{d_m} sec. Each signal is generated by correlating the received signal with the local replica signal at a different code-delay shift. The correlation is calculated by the FFT/IFFT approach. Thus, the generation of each signal requires 3 $N_{T_{dm}} \log_2(N_{T_{dm}})$ computations, where $N_{T_{dm}}$ is the number of samples in T_{d_m} sec.
2. The carrier-tracking module generates output every T_{d_m} sec., but its EKF updates its estimates every T_{car_u} sec. If $T_{car_u} > T_{d_m}$ and the EKF does not update its estimates at the mth interval, the EKF propagates the estimates of the $(m-1)$th interval to the mth interval. The propagation is done using (4.65 to 4.67); this requires three additions and four multiplications. If the EKF is updated, then the computations are as follows. The phase, Doppler shift, and Doppler rate are calculated at the start of the T_{car_u} interval, and the average phase, Doppler shift, and Doppler rate are propagated to the middle of the T_{car_u} interval; this requires six additions and eight multiplications. The measurement, the expected measurement, and the linearization are calculated; the number of computations is linearly proportional to the number of data bits in T_{car_u} sec. The EKF equations are calculated; the number of computations depends on the EKF type. The phase, Doppler shift, and Doppler rate are updated; this requires three additions and three multiplications.
3. The code-tracking module generates output every T_{d_m} sec., but its EKF updates its estimates every T_{code_v} sec. Every T_{d_m} sec., the length of the mth interval, taking into account the Doppler effect on the code length, and the start of the $(m+1)$th interval are estimated. If the EKF is updated, then the following additional computations are needed. The measurement, the expected measurement, and the linearization are calculated; the number of computations is proportional to the number of data bits in T_{code_v} sec. The EKF equations are calculated; the number of computations depends on the EKF type.
4. The navigation message decoding module performs three main tasks: identifying the preamble, identifying the subframe ID, and decoding the navigation message. The most computation is done in the calculation of the parity bits. This is done for every word (i.e., every 30 data bits or 0.6 sec.) if the word has not been decoded.

5.9 Simulation and Results

The developed algorithms are demonstrated using simulated GPS signals with C/A codes and using TCXO and OXO clocks. Settings similar to those described in Sections 3.7 and 4.10 are used here. In addition, the initial code-delay error is modeled to take a value between $(0,\ 1/f_s)$. This value is obtained by generating a uniformly distributed RV between $(0,\ \tau_{Precision}\ f_s)$. Then, the result is multiplied by $1/(\tau_{Precision}\ f_s)$, where $\tau_{Precision}$ is an integer constant. The process noise for the code and carrier tracking is set based on the clock type.

The tracking modules are tested for C/N_0 between 15 and 22 dB-Hz. For each test, the modules run for 900 seconds (15 minutes). The variances of the estimation errors for the tracked parameters are calculated for each test; the calculation is done using the estimation error in each step through out the 900 seconds. Since it is difficult to run each test for a large number of trials to obtain accurate statistics, the following results can be considered an indication of the expected performance of the tracking modules.

The algorithm that decodes the navigation message is implemented as described in Section 5.5. For test purposes, the structure of the navigation message is generated to simulate the structure of the real navigation message. This is done by filling the start of each subframe with the preamble, and filling bits 50 to 52 with each subframe ID. Words 3 to 10 of subframes 1 to 3 are generated with repeated data. The data are generated randomly with a probability of data transition equal to 0.5.

In all the runs, for all the tested C/N_0, the EKFs kept lock on the code, Doppler shift, and Doppler rate when using both the TCXO and OXO clocks. In the phase tracking results, however, there were large performance differences with regard to cycle slipping between the TCXO and OXO clocks. When using a TCXO clock with carrier integration of one data bit interval, the EKF did not encounter any cycle slips for the 22 dB-Hz signal during the 900 seconds. However, for C/N_0 below 22 dB-Hz, cycle slips were encountered, and their number increased as the C/N_0 decreased. When using an OXO clock with carrier integration equal to one data bit interval, the EKF sometimes encountered a few cycle slips at the beginning of its operation for C/N_0 between 15 and 18 dB-Hz, but once the EKF was converged, it either did not encounter any cycle slips during the remaining testing time, or it encountered a very few half-cycle slips. Above 18 dB-Hz, no cycle slips were encountered during the 900 seconds.

Four factors affect the accuracy of the code tracking: the clock type, the integration time, the code-delay separation between the early and late signals, and the EKF type (i.e., first- or second-order EKF).

The code-tracking results are as follows. Figure 5.3 shows the standard deviation, σ_{τ_e} in nanoseconds of the code-delay tracking error versus C/N_0 for different code integration times (Figure 5.4 shows σ_{τ_e} in meters). These results

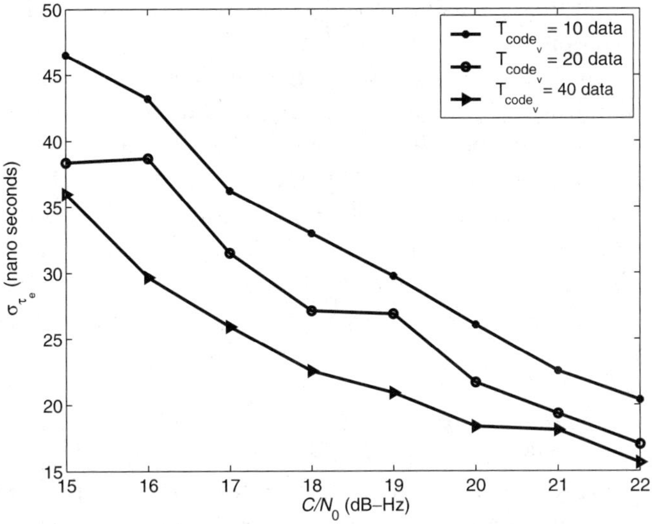

Figure 5.3 Standard deviation of the code-delay error in nanoseconds, using a TCXO clock, GS second-order EKF. $\Delta = 1/f_s$. $T_{code_v} = T_{I_{Max}} = T_{tmax} = T_{tmin} = 10, 20$, and 40 data bit intervals. $T_{I_{Min}} = 1$ data bit interval.

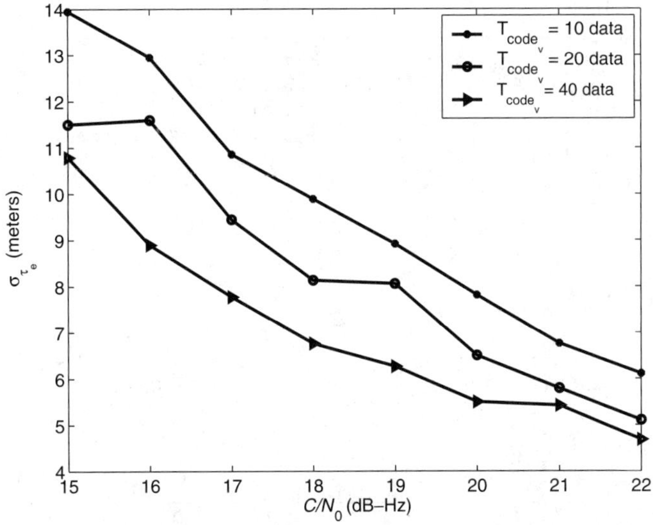

Figure 5.4 Standard deviation of the code-delay error in meters, using a TCXO clock, GS second-order EKF. $\Delta = 1/f_s$. $T_{code_v} = T_{I_{Max}} = T_{tmax} = T_{tmin} = 10, 20$, and 40 data bit intervals. $T_{I_{Min}} = 1$ data bit interval.

are generated using a GS EKF for the code tracking, with $\Delta = 1/f_s$. The code integration length changes adaptively as described in Section 5.6.1. The maximum coherent integration $T_{I_{Max}}$ and the total integrations T_{tmax} and T_{tmin} are set to be equal to each other; $T_{I_{Min}}$ is set equal to one data bit interval. Thus, the code integration length T_{code_v} has the same length throughout the tracking steps, while the coherent integration changes based on the correctly decoded data. Three different total integrations were tested for 10, 20, and 40 data bit intervals. These tests used a first-order square root EKF for the carrier tracking, with $T_{I_{cMax}}$ and $T_{I_{cMin}}$ equal to one data bit interval. Thus, T_{car_u} is equal to one data bit interval throughout the tracking steps, and $\alpha = 40$ Hz/s. A TCXO clock is used.

Figure 5.5 shows σ_{τ_e} in nanoseconds versus C/N_0 (Figure 5.6 shows σ_{τ_e} in meters), using three different settings for the code-delay separation, Δ. These settings are $2/f_s$, $1/f_s$, and both $1/f_s$ and $2/f_s$, as described in Section 5.6.2. Using the two code-delay separations at the same time is tested using two EKFs, as described in (5.123), and one EKF, as described in (5.124) to (5.127). The maximum coherent integration and the total integration are set equal to 20 data bit intervals. The other settings used to generate these results are the same as those used for the results in Figure 5.3. As shown, using Δ of $1/f_s$ or both $1/f_s$ and $2/f_s$ with two EKFs gave better performance as compared to using Δ of $2/f_s$. However, using neither Δ of $1/f_s$ nor the two delay separations together with two

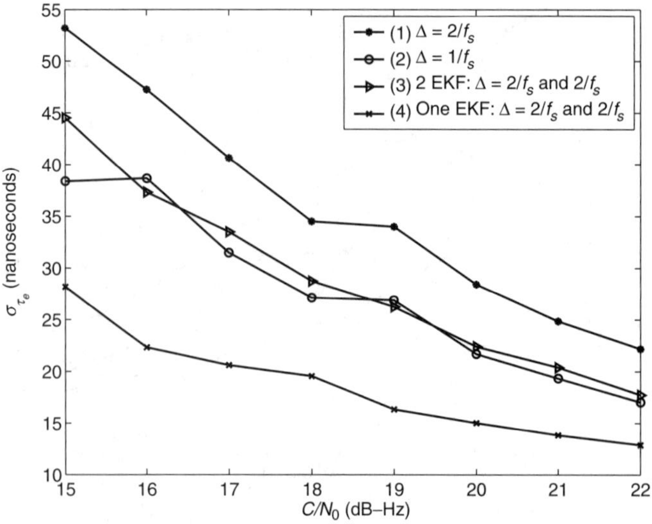

Figure 5.5 Standard deviation of the code-delay error in nanoseconds, using a TCXO clock, of Δ of (1) $2/f_s$; (2) $1/f_s$; (3) $1/f_s$ and $2/f_s$ using two EKFs; and (4) $1/f_s$ and $2/f_s$ using one EKF. $T_{code_v} = T_{I_{Max}} = T_{tmax} = T_{tmin} = 20$ data bit intervals. $T_{I_{Min}} = 1$ data bit interval.

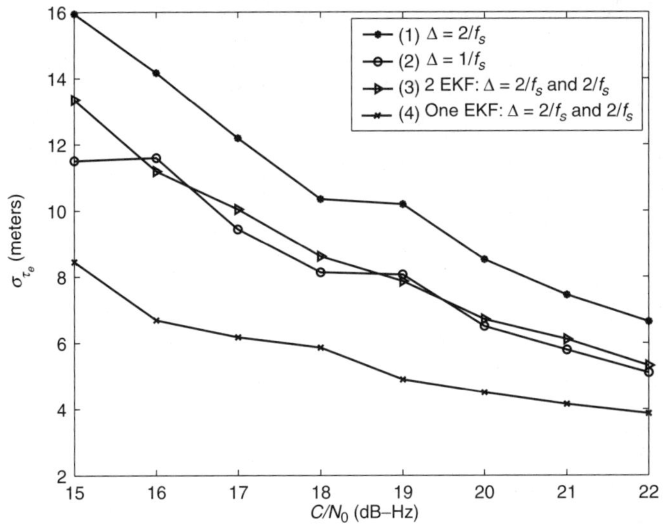

Figure 5.6 Standard deviation of the code-delay error in meters, using a TCXO clock, of Δ of (1) $2/f_s$; (2) $1/f_s$; (3) $1/f_s$ and $2/f_s$ using two EKFs; and (4) $1/f_s$ and $2/f_s$ using one EKF. $T_{code_v} = T_{I_{Max}} = T_{tmax} = T_{tmin} = 20$ data bit intervals. $T_{I_{Min}} = 1$ data bit interval.

EKFs gave apparent better performance, but using the two code-delay separations together with one EKF gave the best code-tracking performance.

A code-delay error of 3 ns causes a 1m position error. Therefore, if other sources of errors are neglected, it is possible to achieve a position accuracy of less than 9m for 15 dB-Hz signals using code integration of 20 data bit intervals and a position accuracy of less than 4m for 22 dB-Hz signals using code integration of 20 data bit intervals. These errors can be reduced further by increasing the code integration length.

The TS EKF for the code tracking was also tested. The results were comparable to the GS EKF. Also, the code tracking was tested using α up to 300 Hz/s. The results were approximately the same as those obtained using α of 40 Hz/s.

The factors affecting the accuracy of the carrier tracking include the clock type, the maximum integration length $T_{I_{cMax}}$, and the EKF type (i.e., first- or second-order square root EKF).

The carrier-tracking results are as follows. Figure 5.7 shows the standard deviation, σ_{f_e}, of the Doppler shift tracking error versus C/N_0 using both TCXO and OXO clocks. For each clock, two different tests were performed. Each test used different $T_{I_{cMax}}$; one test used $T_{I_{cMax}}$ equal to one data bit interval, while the other test used $T_{I_{cMax}}$ equal to five data bit intervals, but both tests used $T_{I_{cMin}}$ equal to one data bit interval. For the case of $T_{I_{cMax}}$ equal to five data bit intervals, σ_{f_e} was calculated for the steps in which T_{car_u} was larger than one data bit interval to show the difference in the performance between using T_{car_u} of one data bit interval

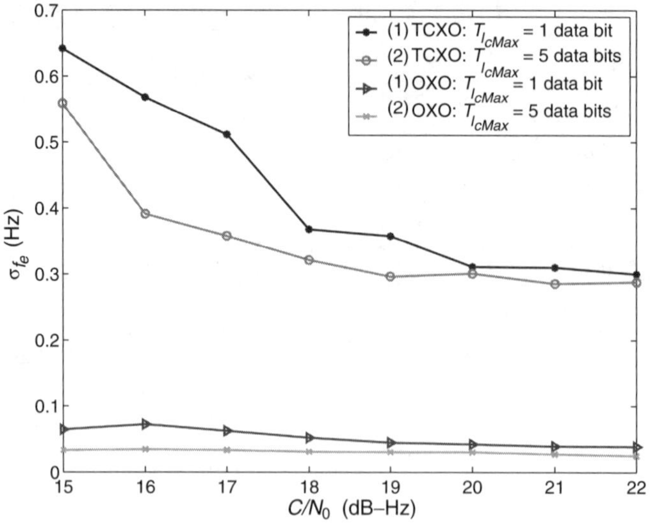

Figure 5.7 Standard deviation of the Doppler shift estimation error, using a first-order square root EKF with (1) $T_{I_{cMax}} = T_{I_{cMin}} = T_{car_u} = 1$ data bit interval, and (2) $T_{I_{cMax}} = 5$ data bit intervals, and $T_{I_{cMin}} = 1$ data bit interval. For the cases of $T_{I_{cMax}} = 5$ data bit intervals, the standard deviations are calculated for the steps in which T_{car_u} is larger than one data bit interval, for both TCXO and OXO clocks.

and of larger than one data bit interval. These results were generated using a first-order square root EKF for the carrier tracking, with $\alpha = 40$ Hz/s. They also used a GS EKF for the code tracking, with $T_{code_v} = T_{I_{Max}} = T_{tmax} = T_{tmin} = 20$ data bit intervals, $T_{I_{Min}} = 1$ data bit interval, and $\Delta = 1/f_s$.

Figure 5.8 shows the standard deviation, σ_{θ_e}, of the phase tracking error versus C/N_0 for the same test cases as those of Figure 5.7. It should be noted that σ_{θ_e} is calculated after removing the effect of cycle slips. Not shown in the figures is the fact that increasing $T_{I_{cMax}}$ caused improvement in the performance of the phase tracking with regard to cycle slipping. Using a second-order square root EKF for the carrier tracking did not cause a large performance difference as compared to using a first-order square root EKF.

Figures 5.9 and 5.10 show σ_{f_e} and σ_{θ_e} versus Doppler rate. The tested Doppler rates ranged from 0 to 300 Hz/s. The results are for C/N_0 of 15 and 18 dB-Hz. These results were generated using a first-order square root EKF for the carrier tracking, with $T_{I_{cMax}}$ equal to one data bit interval, while the other settings are the same as those used in generating the previous two figures.

The algorithm that decodes the navigation message was implemented in all of the test runs. The average number of subframes it took the algorithm to find the preamble and to identify the subframes IDs is shown in Figure 5.11.

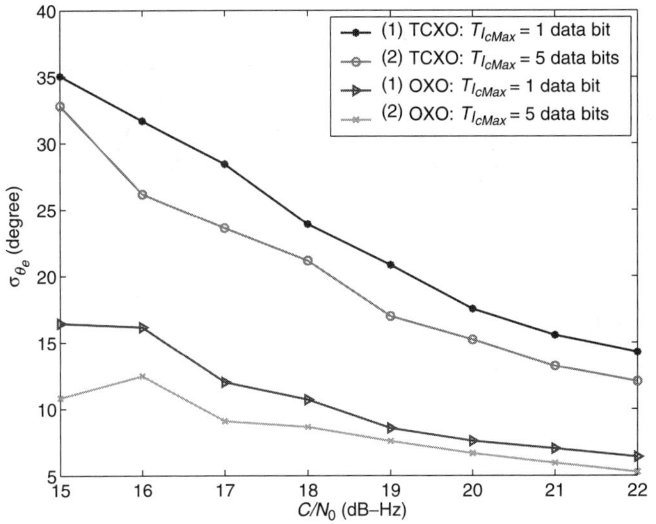

Figure 5.8 Standard deviation of carrier-phase error using a first-order square root EKF with (1) $T_{I_{cMax}} = T_{I_{cMin}} = T_{car_u} = 1$ data bit interval, and (2) $T_{I_{cMax}} = 5$ data bit intervals, and $T_{I_{cMin}} = 1$ data bit interval. For the cases of $T_{I_{cMax}} = 5$ data bit intervals, the standard deviations are calculated for the steps in which T_{car_u} is larger than one data bit interval for both TCXO and OXO clocks.

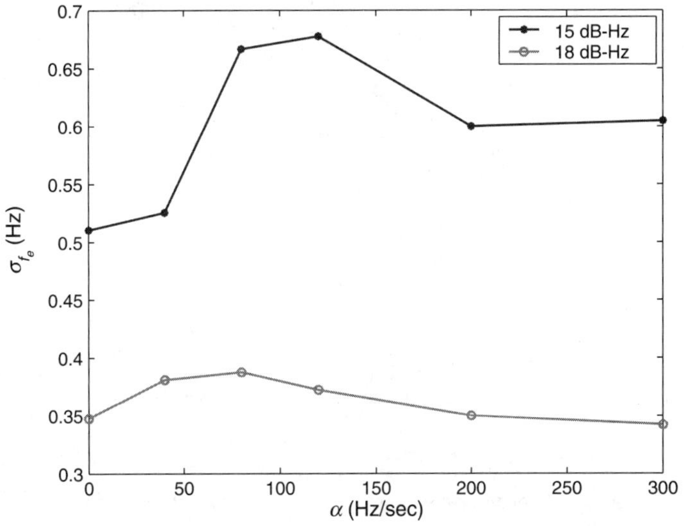

Figure 5.9 α versus standard deviation of the Doppler shift estimation error using a TCXO clock and first-order square root EKF with $T_{I_{cMax}} = 1$ data bit interval.

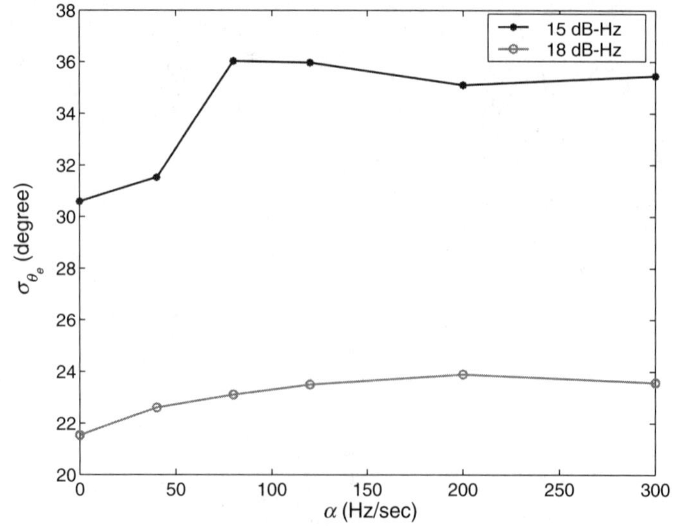

Figure 5.10 α versus standard deviation of the phase estimation error using TCXO a clock and first-order square root EKF, with $T_{I_{cMax}} = 1$ data bit interval.

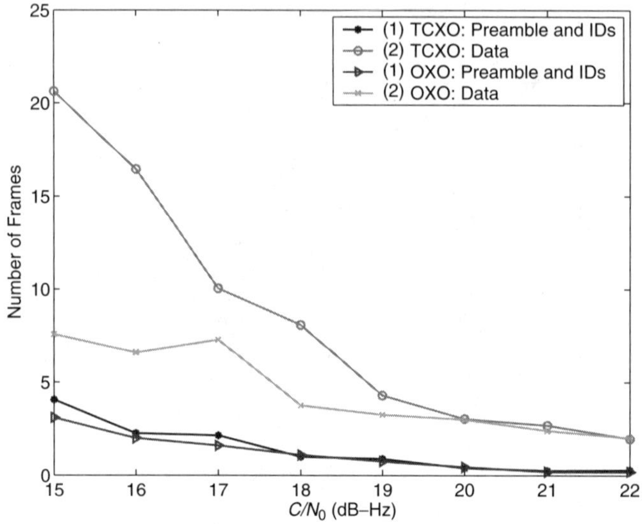

Figure 5.11 Average number of frames versus C/N_0 needed to (1) identify the preamble and the subframe's ID, and (2) decode the whole repeated subframes, using both TCXO and OXO clocks.

Each subframe duration is 30 seconds. The results are for both the TCXO and OXO clocks. Also shown in the same figure is the average number of frames it took the algorithm to decode the data bit of words 3 to 10 of subframes 1 to 3. The total number of these words is 24. These are the words that repeat every 30 seconds. Those results were generated using the following settings in the navigation message decoding algorithm: $N_{min} = 3$, $\gamma_{preamble} = 0$, $ID_{th} = 4$, and $\gamma_{level} = 0.55$. It should be noted that in the real navigation message, words 1 and 2 of subframes 1 to 3 can be deduced and used to help in the tracking. The same algorithm can be used to decode subframes 4 and 5, which repeat every 12.5 minutes. Decoding the whole navigation message will enhance the tracking performance.

Figures 5.12 to 5.19 show the tracking-error histories, along with the square root of the EKF covariance, for the code delay and Doppler shift during 30 minutes of operation (i.e., 1,800 sec.). The results are for C/N_0 of 15 and 18 dB-Hz. Figures 5.12 to 5.15 were generated using a code-delay separation $\Delta = 1/f_s$, while Figures 5.16 to 5.19 were generated using two code-delay separations at the same time with only one EKF; $\Delta = 1/f_s$ and $2/f_s$. Those results were generated with the following settings: TCXO clocks, a GS EKF for the code tracking with $T_{I_{Max}} = T_{tmax} = T_{tmin} = 20$ data bit intervals, $T_{I_{Min}} = 1$ data bit interval, a first-order square root EKF for the carrier tracking with $T_{I_{cMax}} = 1$ data bit interval, and $\alpha = 40$ Hz/s.

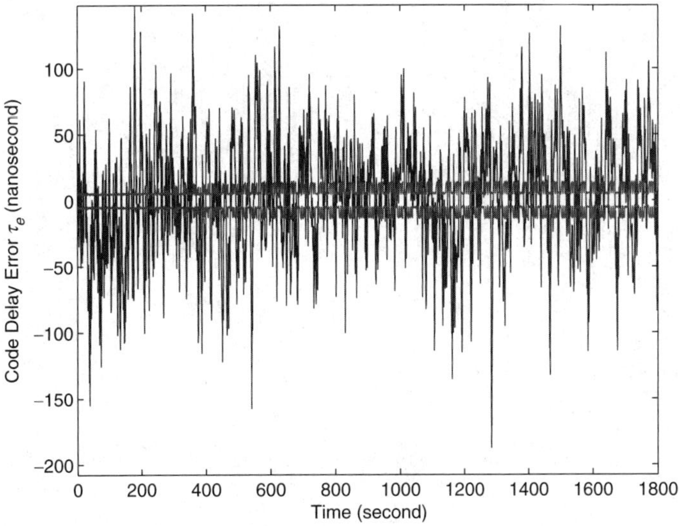

Figure 5.12 Code-delay error history for 30 minutes of operation and the square root of the EKF covariance for a 15-dB-Hz signal using a TCXO clock; $\Delta = 1/f_s$, $\alpha = 40$ Hz/s, $T_{code_v} = 20$ data bit intervals, and $T_{I_{cMax}} = 1$ data bit interval.

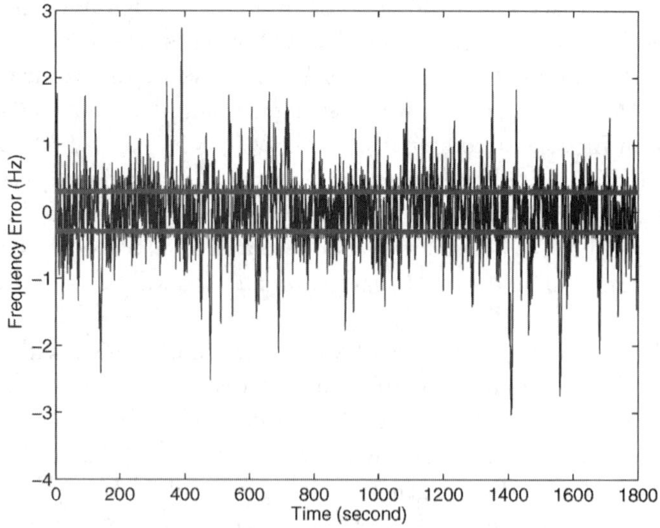

Figure 5.13 Doppler shift error history for 30 minutes of operation and the square root of the EKF covariance for a 15-dB-Hz signal using a TCXO clock; $\Delta = 1/f_s$, $\alpha = 40$ Hz/s, $T_{code_v} = 20$ data bit intervals, and $T_{I_{cMax}} = 1$ data bit interval.

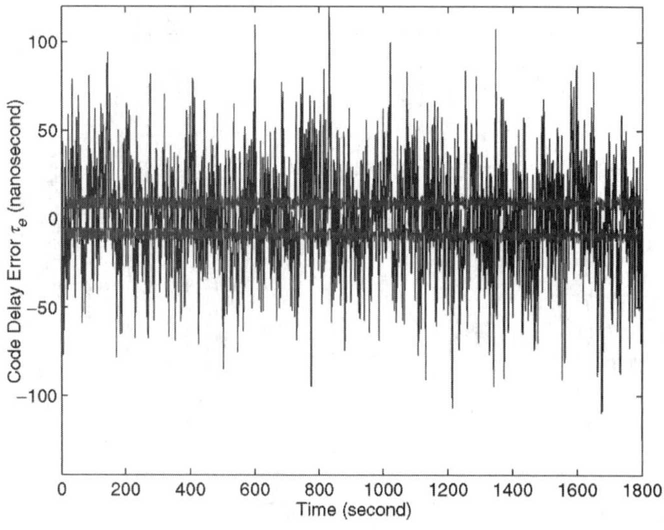

Figure 5.14 Code-delay error history for 30 minutes of operation and the square root of the EKF covariance for an 18-dB-Hz signal using a TCXO clock; $\Delta = 1/f_s$, $\alpha = 40$ Hz/s, $T_{code_v} = 20$ data bit intervals, and $T_{I_{cMax}} = 1$ data bit interval.

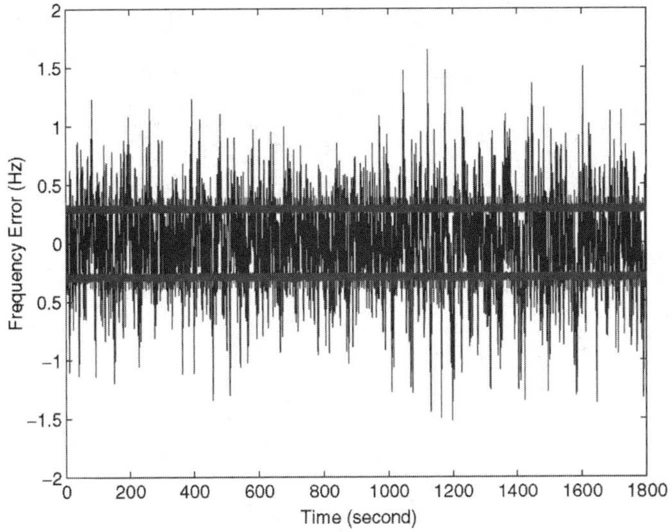

Figure 5.15 Doppler shift error history for 30 minutes of operation and the square root of the EKF covariance for an 18-dB-Hz signal using a TCXO clock; $\Delta = 1/f_s$, $\alpha = 40$ Hz/s, $T_{code_v} = 20$ data bit intervals, and $T_{I_{cMax}} = 1$ data bit interval.

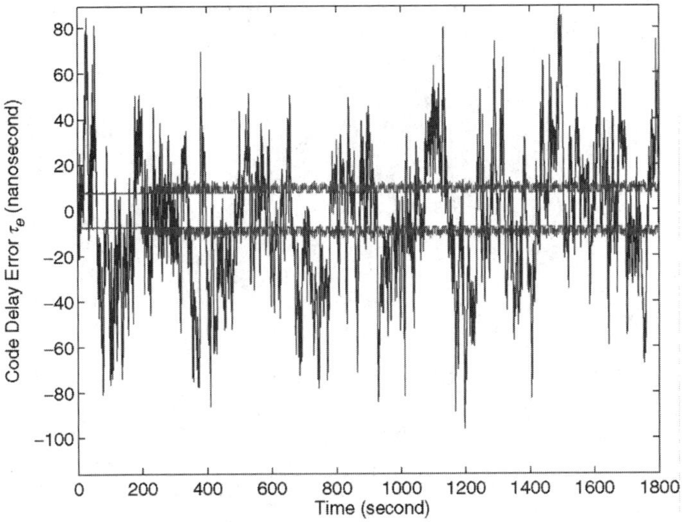

Figure 5.16 Code-delay error history for 30 minutes of operation and the square root of the EKF covariance for a 15-dB-Hz signal using a TCXO clock; $\Delta = 1/f_s$ and $2/f_s$ with one EKF, $\alpha = 40$ Hz/s, $T_{code_v} = 20$ data bit intervals, and $T_{I_{cMax}} = 1$ data bit interval.

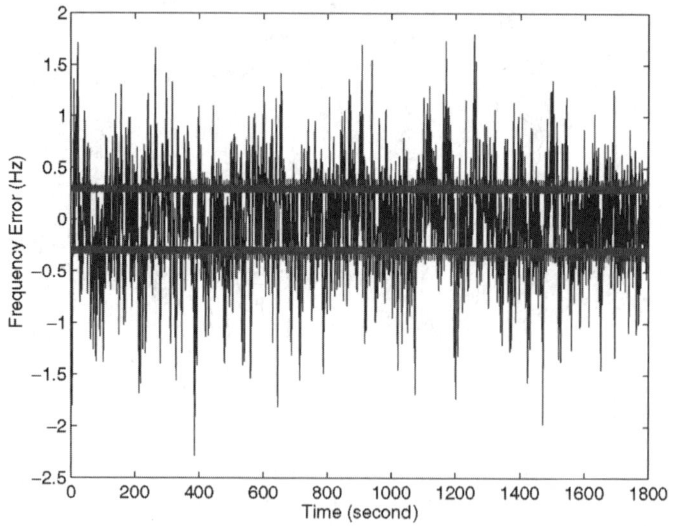

Figure 5.17 Doppler shift error history for 30 minutes of operation and the square root of the EKF covariance for a 15-dB-Hz signal using a TCXO clock; $\Delta = 1/f_s$ and $2/f_s$ with one EKF, $\alpha = 40$ Hz/s, $T_{code_v} = 20$ data bit intervals, and $T_{I_{cMax}} = 1$ data bit interval.

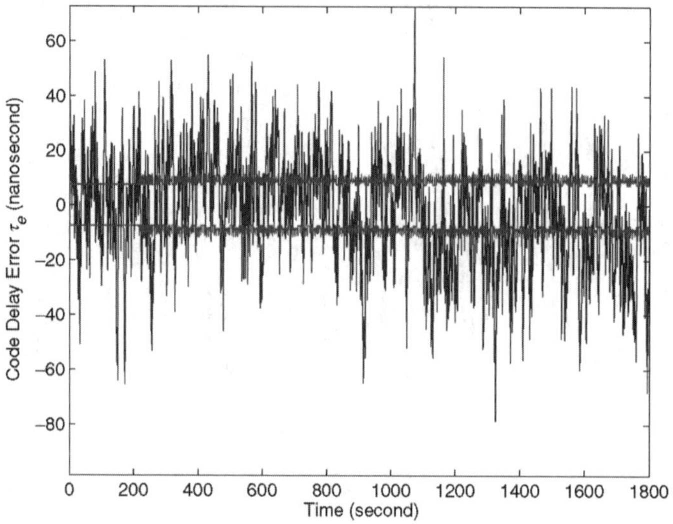

Figure 5.18 Code-delay error history for 30 minutes of operation and the square root of the EKF covariance for an 18-dB-Hz signal using a TCXO clock; $\Delta = 1/f_s$ and $2/f_s$ with one EKF, $\alpha = 40$ Hz/s, $T_{code_v} = 20$ data bit intervals, and $T_{I_{cMax}} = 1$ data bit interval.

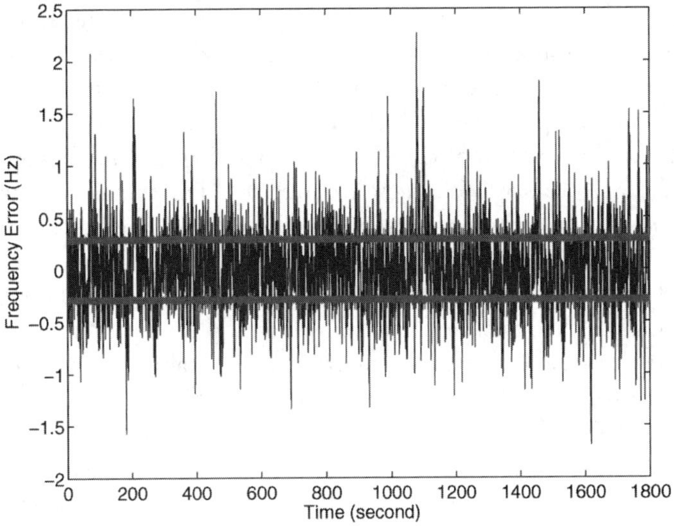

Figure 5.19 Doppler shift error history for 30 minutes of operation and the square root of the EKF covariance for an 18-dB-Hz signal using a TCXO clock; $\Delta = 1/f_s$ and $2/f_s$ with one EKF, $\alpha = 40$ Hz/s, $T_{code_v} = 20$ data bit intervals, and $T_{I_{cMax}} = 1$ data bit interval.

As shown in the figures, the EKFs kept lock on the code and Doppler shift throughout the 30 minutes of operation; however, using the two code-delay separations gave better code-tracking performance. The standard deviations of the code and carrier tracking over the 30 minutes are shown in Table 5.1.

5.10 Summary and Conclusions

Code- and carrier-tracking algorithms are developed in this chapter to work with very weak signals and under different dynamic conditions. The algorithms are based on EKF approaches. Different EKF designs are introduced that include first-order, second-order, and square root EKFs. A first-order EKF requires the

Table 5.1
The Standard Deviations of the Code and Carrier Tracking for C/N_0 of 15 and 18 dB-Hz

C/N_0 (dB-Hz)	Δ	σ_τ (ns)	σ_{f_e} (Hz)	$\sigma^\circ_{\theta_e}$
15	$1/f_s$	38	0.58	32
15	$1/f_s$ and $2/f_s$	29	0.48	31
18	$1/f_s$	30	0.39	22
18	$1/f_s$ and $2/f_s$	20	0.39	22

least processing but can become unstable due to numerical precision problems that can lead the covariance matrix to have negative eigenvalues. The square root EKF updates the covariance matrix in a way that preserves its symmetric- and positive-definiteness properties, which in turn guarantee that the eigenvalues are positive. Thus, the square root EKF is more stable than the first-order EKF. The second-order EKF can provide more accurate tracking because it makes use of higher orders in the Taylor series expansion in deriving the state and covariance update equations.

In addition, a navigation message decoding algorithm is developed. The algorithm utilizes the message structure to identify the preamble and subframe ID, then to decode the rest of the message. This enables the decoding of the navigation message even for very weak C/N_0 signals that have a high BER. Also, some approaches are developed to maximize the time of the tracking. The time of the transmission of the signal from the satellite to the receiver can be calculated using the output of the tracking modules; and the ephemeris, which is extracted from the navigation message. So, the pseudorange and the navigation solution can be calculated as in [4].

The results have shown the ability of these algorithms to work with signals as low as 15 dB-Hz and with a Doppler rate as high as 300 Hz/s, using TCXO clocks. For a 15-dB-Hz signal, the Doppler shift can be resolved to an accuracy of less than 0.5 Hz with an integration of one data bit interval, while the code delay can be resolved to an accuracy of less than 25 ns with an integration of 20 data bit intervals, which corresponds to less than a 9m position accuracy. The navigation message can also be decoded; thus, it is possible to calculate the navigation solution for the 15-dB-Hz signal.

The algorithms could also work with weaker signals and under higher dynamic conditions; however, it is very rare for a signal to be attenuated to below 18 dB-Hz. Also, the Doppler rate of 300 Hz/s corresponds to 6g of acceleration, which is very high and not likely to be encountered in the applications for which this work is intended.

References

[1] Maybeck, P. S., *Stochastic Models, Estimation, and Control*, Vol. 2, Burlington, MA: Academic Press, 1982.

[2] Maybeck, P. S., *Stochastic Models, Estimation, and Control*, Vol. 1, Burlington, MA: Academic Press, 1979.

[3] NAVSTAR GPS Space Segment, Navigation User Interface Control Document (ICD-GPS-200).

[4] Parkinson, B., and J. Spilker. *Global Positioning System: Theory and Applications*, Washington, D.C.: AIAA, 1996.

6

Summary and Conclusions

6.1 Introduction

This chapter presents an overall summary of the algorithms presented in this book and their performance. It also presents a performance comparison between some conventional acquisition and tracking algorithms and the novel algorithms in this book. This chapter is organized as follows. Section 6.2 summarizes all the presented algorithms and outlines the calculation of the navigation solution. Section 6.3 summarizes the acquisition functionalities and presents a performance comparison between aided and unaided conventional acquisition algorithms and the CCMDB and MDBZP algorithms presented in this book. Section 6.4 summarizes the fine acquisition and tracking functionalities and presents a performance comparison between aided and unaided PLL, FLL, DLL, and the EKF-based tracking algorithms presented in this book. Section 6.5 outlines the acquisition, fine acquisition, code- and carrier-tracking, and navigation message decoding of the new L2C and L5 civil signals.

6.2 Summary of the Algorithms

Fifteen novel GPS receiver algorithms are introduced in this book. The algorithms are designed to work with very weak signals under low and high dynamic conditions. The processing and memory requirements have been considered in the design of the algorithms to allow them to fit the limited resources of applications such as the positioning of wireless devices. The algorithms are divided into seven modules, and they cover all of the main receiver functions. The modules include signal generator, acquisition, fine acquisition, bit synchronization and data detection, code and carrier tracking, and navigation message decoding. The modules are integrated and use each others' output. Each module is designed to work independently of the other modules to allow the possibility of integrating each module into any GPS receiver. Figure 6.1 shows an overview of the interaction between different modules. Detailed interactions between different modules are discussed in Sections 4.3 and 5.2.

The developed algorithms are as follows: two algorithms for the acquisition of weak signals, the CCMDB and MDBZP; one algorithm for the acquisition of weak signals in the presence of strong interfering signals; one algorithm for high dynamic acquisition; fine acquisition and high dynamic fine acquisition algorithms; a bit synchronization and data detection algorithm; code- and carrier-tracking algorithms that work under low and high dynamic conditions; one algorithm to detect and correct large carrier-tracking errors; one algorithm to detect and correct large code-tracking errors; one algorithm to deal with large sudden changes in the receiver dynamics; preamble identification, subframe identification, and navigation message decoding algorithms.

The time of travel of the signal is calculated from the code-delay estimate. The ephemeris is extracted from the decoded navigation message. Thus, the pseudorange and, consequently, the navigation solution are calculated as discussed in Section 1.3.

All the algorithms have been demonstrated using simulated GPS C/A codes and both OXO and TCXO clocks. The results have shown the ability of these algorithms to work reliably with signals as low as 10 dB-Hz and with Doppler rates over 300 Hz/s, which corresponds to 6g of receiver acceleration. The algorithms

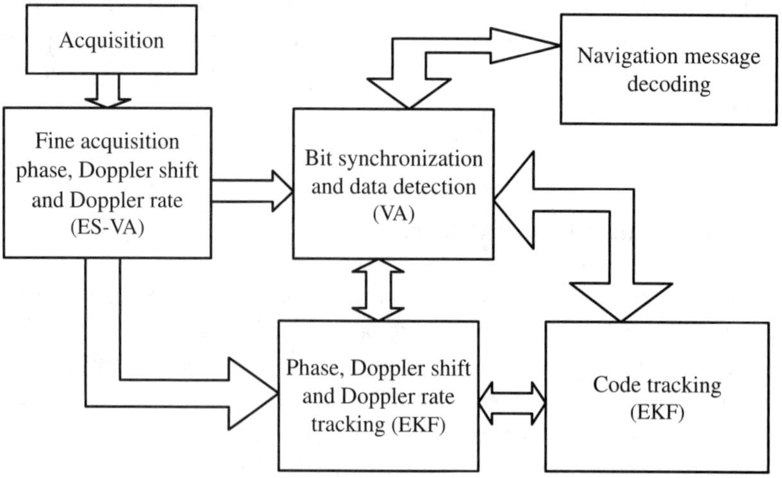

Figure 6.1 Overview of the interaction between different receiver modules.

can be implemented in real GPS software receivers. They can fit a variety of applications.

6.3 Summary of the Acquisition Algorithms and Their Performance

6.3.1 Acquisition Algorithm Functionalities

The CCMDB and MDBZP acquisition algorithms use coherent integration that is a multiple of one data bit interval with no assisting information. The two algorithms differ in the method used to calculate the coherent integration: the CCMDB algorithm uses circular correlation, and the MDBZP algorithm uses a modified version of the DBZP. The problems associated with long integration have been circumvented. The most likely data bit combination is estimated in each coherent integration interval, and then it is used to remove the effect of the data bit signs before incorporating the coherent integration into a longer incoherent accumulation. The increased processing due to the increased number of Doppler bins associated with the increase in the coherent integration length is circumvented by starting the acquisition process with a small integration length, performing a few accumulations, concluding that the most likely incorrect bins are those that accumulate the minimum powers and eliminating them, and then increasing the coherent integration without considering the eliminated Doppler bins. Thus, the maximum number of Doppler bins can be controlled when using any coherent integration length. This process can be repeated several times; thus, long coherent integration can be used without adding processing overhead.

In addition, the inability of the DBZP to use long coherent and incoherent integrations due to the use of onc replica code, which is not compensated in length by the Doppler effect, has been dealt with in a modified version, MDBZP. This is done by using a small number of replica code versions. Each version is compensated in length by the effect of a different Doppler shift. Those Doppler shifts are chosen to cover all of the Doppler range (i.e., they are uniformly spaced over the whole Doppler range). The change in the relative code delay between the received and the replica code is also compensated for in the MDBZP.

The algorithm that acquires weak signals in the presence of strong interfering signals utilizes the cross-correlation properties of the strong signals. The high dynamic acquisition is achieved by continuously estimating the most likely closest bin to the current Doppler shift, where the powers of the closest bins are accumulated together. The output of the acquisition module includes approximate Doppler shift, Doppler rate, code delay, and bit edge position. The probabilities of false alarm and detection are derived, taking into consideration all of the approaches used in those algorithms.

6.3.2 Acquisition Performance

There are two types of conventional acquisition algorithms: unaided and aided acquisition. An unaided acquisition algorithm can use a maximum of 20-ms coherent integration [1]. This is done by calculating 20 different integrations. Each integration starts at a possible bit edge position. Weak signals can be acquired by incoherently accumulating consecutive coherent integration results. However, more data will be required to acquire a weak signal compared to the data needed by either of the developed CCMDB or MDBZP algorithms. An aided acquisition algorithm can have aiding information, such as the bit edge position, data bit values, and approximate Doppler shift. Therefore, long coherent and incoherent integrations can be used.

The amount of data needed to acquire a signal is determined from the probabilities of false alarm, p_f, and detection, p_d. Figure 6.2 shows a typical PDF of p_d and p_f. The area of intersection between the PDF of the p_d and p_f is an indication of the possible p_d and p_f values that can be achieved. The acquisition threshold is chosen to give acceptable values for both the p_d and p_f. When the intersection area becomes relatively large, the amount of data used to generate the PDF becomes insufficient to acquire a signal. There is, however, another factor that affects the calculation of the p_f. This is the number of cells, N_{Result}, that are considered in the acquisition (i.e., the number of possible Doppler shifts, the number of possible code delays, and the number of possible bit edge positions). As N_{Result} increases, more data will be needed to acquire a signal because p_f must be scaled by N_{Result}.

Assuming that no data bit combinations are considered, the probability of false alarm using L incoherent accumulations is

$$f_S(s) = N_{Result} \int_{\gamma}^{\infty} \frac{1}{2^L} \frac{s^{L-1}}{(L-1)!} e^{-\frac{s}{2}} ds \tag{6.1}$$

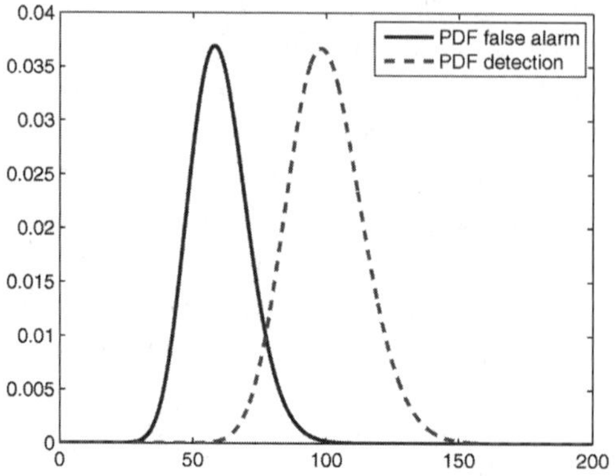

Figure 6.2 PDF of the p_f and p_d.

where $N_{Result} = N_{f_d} N_\tau N_b$. N_{f_d} is the number of Doppler bins, N_τ is the number of code delays, and N_b is the number of bit edge positions. N_{f_d} increases with the increase in the coherent integration length. The PDF of the p_f is independent of the coherent integration length; it depends on the number of incoherent accumulations.

Figure 6.3 plots the PDF of this p_f using 200, 100, and 50 as the number of incoherent accumulations. Assuming that the total amount of data used is 4 seconds, the number of accumulations correspond to PITs of 20, 40, and 80 ms, respectively. If the data bit combinations are considered and the combination that generates the maximum power is taken from each cell independently of other cells as described in Section 3.2.7 and (3.16), then the PDF of p_f is determined as described in Chapter 3, Section 3.6.4, and by (3.96) or (3.97). Figure 6.4 plots the PDF of this p_f using 200, 100, and 50 as the number of incoherent accumulations, assuming the total amount of data used is 4 seconds.

Figures 6.3 and 6.4 show that when using a PIT larger than 20 ms (i.e., when there is more than one possible data bit combination), the PDF is shifted to the right when considering the possible data bit combinations as compared to not considering them. This indicates that when the possible data bit combinations are considered, as described in Section 3.2.7, larger threshold will be required to get the same p_f value compared to not considering the data bit combination. However, the amount of data needed to acquire a signal is also determined from the probability of detection.

Assuming the data bits are known, the probability of detection using a T_I coherent integration and L incoherent accumulations is

$$p_d = \int_\gamma^\infty \frac{1}{2}\left(\frac{s}{A^2 L^2}\right)^{\frac{1}{2}(L-1)} e^{-\frac{1}{2}(s+A^2L^2)}\, I_{L-1}(\sqrt{s}\,AL)\, ds \qquad (6.2)$$

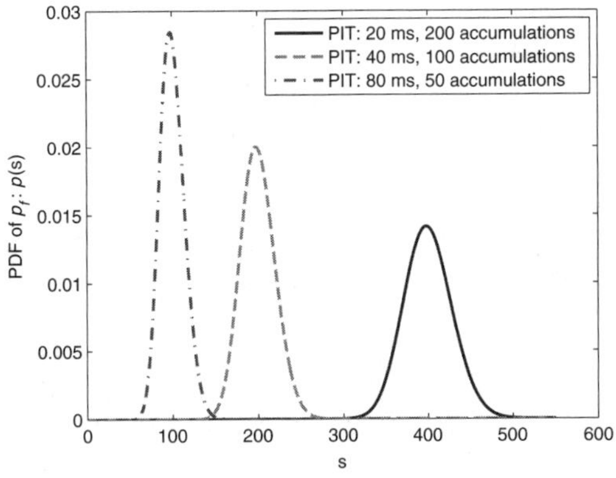

Figure 6.3 PDF of the p_f using 200, 100, and 50 accumulations and 4 seconds of total data (i.e., using T_I of 20, 40, and 80 ms, respectively) and assuming that the data bit combinations are not considered.

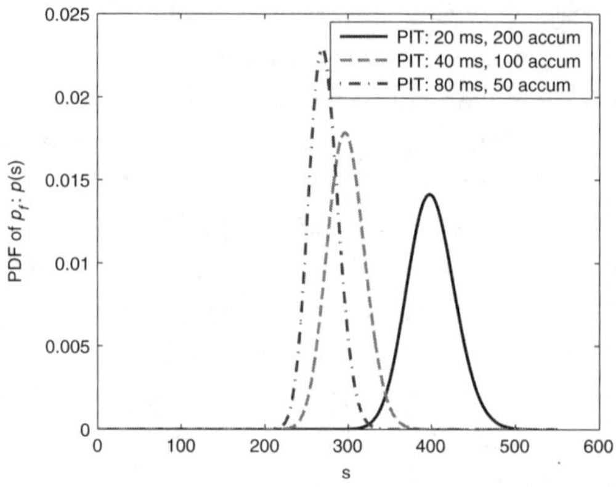

Figure 6.4 PDF of the p_f using 200, 100, and 50 accumulations and 4 seconds of total data (i.e., using T_I of 20, 40, and 80 ms, respectively) and assuming that the data bit combination that generates the maximum power is taken as the likely data bit combination.

where A is the signal amplitude, which is a function of the C/N_0 and T_I. Figure 6.5 plots the PDF of the p_d for C/N_0 of 15 dB-Hz and T_I of 20, 40, and 80 ms. The total amount of data used is 4 seconds. If the data bits are unknown, then the PDF of p_d is determined as described in Chapter 3, Section 3.6.5, and by (3.116). Figure 6.6 plots the PDF of the p_d in this case for C/N_0 of 15 dB-Hz and T_I of 20, 40, and 80 ms. The total amount of data used is also 4 seconds.

Figures 6.5 and 6.6 show that when using a PIT larger than 20 ms (i.e., when there is more than one possible data bit combination), the PDF is shifted to the right when considering the possible data bit combinations as compared to assuming that the data bits are known. This is also the case with the PDF of the p_f,

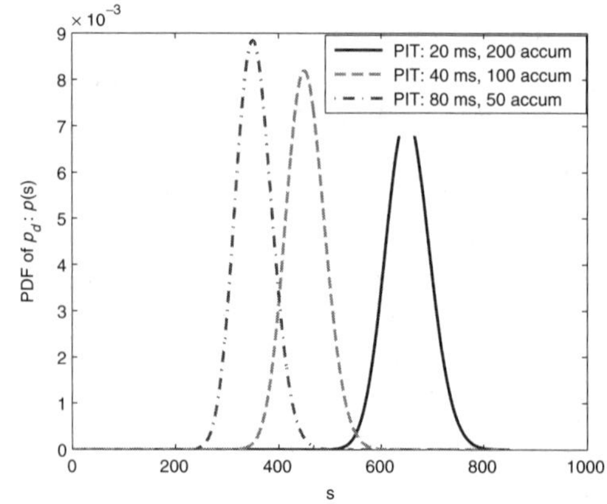

Figure 6.5 PDF of the p_d using T_I of 20, 40, and 80 ms and 4 seconds of total data (i.e., using 200, 100, and 50 accumulations, respectively) and assuming that the data bit values are known.

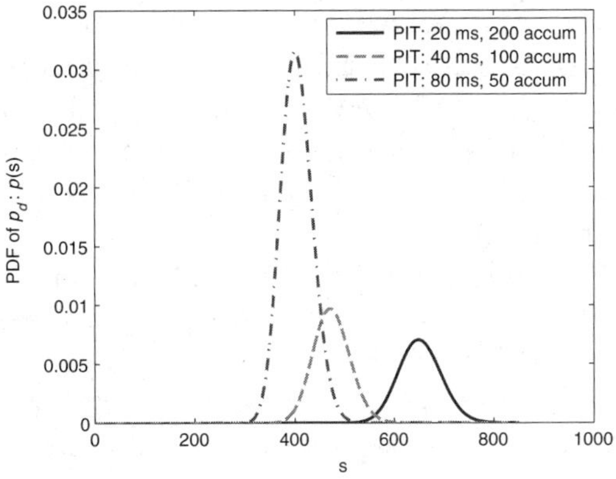

Figure 6.6 PDF of the p_d using T_I of 20, 40, and 80 ms and 4 sec. of total data (i.e., using 200, 100, and 50 accumulations, respectively) and assuming that the data bit combination that generates the maximum power is taken as the likely data bit combination.

which indicates that when the possible data bit combinations are considered, as described in Section 3.2.7, a larger threshold will be required to get the same p_d and p_f values compared to not considering the data bit combinations.

Figures 6.3 to 6.6 show that the PDF of p_f is shifted more to the right compared to the PDF of p_d. Therefore, it is expected that the amount of data needed to acquire a signal when considering the data bit combination that generates the maximum power will be larger than the amount of data needed to acquire the same signal in the case when the data bits are known. It should be noted however that the p_f must be scaled by the number of cells, $N_{Results}$.

To give an illustration of the performance difference when using different approaches for acquisition, four tests were conducted for each C/N_0 of 10, 15, and 17 dB-Hz. Each test used a different acquisition technique. The four acquisition techniques were as follows:

1. *A coherent integration of 20 ms, followed by incoherent accumulations.* 20 possible bit edge positions were considered.
2. *The MDBZP presented in Chapter 3, Section 3.3.* An 80-ms coherent integration was used, followed by incoherent accumulation, as described in the test in Section 3.7, and 10 possible bit edges were considered.
3. *The CCMDB algorithm presented in Chapter 3, Section 3.2.* Coherent integrations of 20 to 120 ms were used, followed by incoherent accumulation, as described in the test in Section 3.7, and four possible bit edges were considered.
4. *Only coherent integration.* Aided information of Doppler shift, code delay, and data bit values were assumed to be available.

In the test of technique (1), the acquisition was run until the correct cell gave a clear peak to indicate a correct acquisition. The results of techniques (2) and (3) are taken from Section 3.7. In the test of technique (4), the acquisition was run with different coherent integration lengths, and the minimum length that gave a clear peak at the correct Doppler shift and code delay was taken as the length needed to acquire the signal.

Figure 6.7 shows the result of the tests. The amount of data needed to acquire a signal is plotted versus the C/N_0. The performance advantage of both the MDBZP and CCMDB compared to the conventional approach of using 20-ms coherent integration followed by incoherent accumulations can be seen. As expected, the availability of aided data reduces the total amount of data needed for acquisition. In reality, however, the aided data are not always available; therefore, a stand-alone GNSS receiver will be needed.

6.4 Summary of the Fine Acquisition and Tracking Algorithms and Their Performance

6.4.1 Fine Acquisition and Tracking Algorithm Functionalities

The fine acquisition algorithm uses a modified version of the VA, the ES-VA. The algorithm works iteratively to provide accurate estimates of the Doppler shift, Doppler rate, and phase, with small processing. Although the acquisition algorithms provide approximate bit edge positions, the bit synchronization is designed to work even if no approximate bit edge position is available. The bit edge position and the data bit sequence are estimated by the VA; EKFs are used in the VA survivor paths to remove the errors of the phase, Doppler shift, and

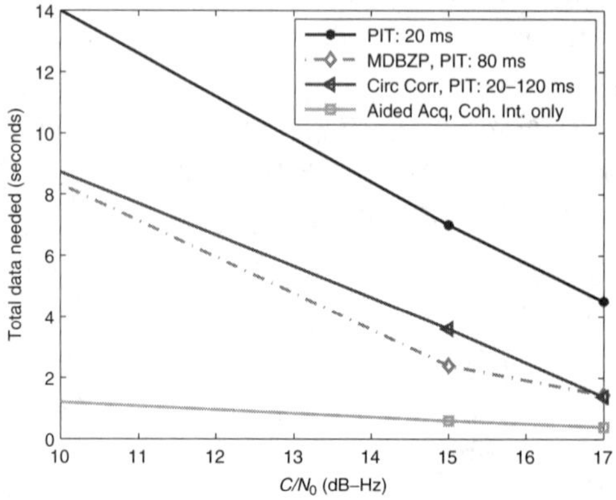

Figure 6.7 Comparison of the amount of data needed for positive acquisition using different acquisition techniques.

Doppler rate. A method that continues to delete the unlikely correct edge positions is introduced; it works until only one edge position is left. This edge position is concluded to be the estimated one.

Several EKF designs are introduced for the code and carrier tracking. The first-order EKF is the simplest, the second-order EKF is more accurate, and the square root EKF is more stable and does not suffer from numerical errors. Several approaches are developed to increase the time to lose look. The carrier-tracking module can adaptively change the integration length based on the correctly decoded data. The code-tracking module has the ability to change the coherent and incoherent integration lengths adaptively, also based on the correctly decoded data. The code tracking can change the code-delay separation, and it has the ability to use more than one separation at the same time. A method is developed to continue to monitor the code- and carrier-tracking performances and to detect large tracking errors. If a large carrier-tracking error is detected, then the EKF for the carrier tracking is deactivated, and a reinitialization module is activated to refine the errors. Following that, the EKF for the carrier tracking is reactivated. Similarly, if a large code-delay error is detected, the EKF for the code tracking is deactivated and an acquisitionlike process is activated. This process searches for the correct code delay using only a small number of delays located around the current code-delay estimate. The detection of large errors avoids the problem of completely losing lock and having to reacquire the signal. Also, an approach is developed to detect large random changes in the receiver velocity or acceleration.

The navigation message decoding algorithms utilize the message structure and the fact that subframes 1 to 3 repeat every 30 seconds, and subframes 4 and 5 repeat every 12.5 minutes. First, the preamble and subframe IDs are identified, and then the rest of the message is decoded. The message decoding of signals with low C/N_0 and high BER is done by adding the signed signal levels of the repeated data. This increases the SNR and decreases the BER and, thus, enables correct data detection. The detected repeated data are used to detect large Doppler shift errors. A large Doppler shift error is concluded if the BER of the newly detected repeated data, which have already been decoded in previous steps, is higher than a certain threshold.

6.4.2 Tracking Performance

Many techniques can be used for the code and carrier tracking [2–7]. The most common technique is the tracking loop. Examples of other tracking techniques include Kalman filtering [3, 4, 6, 8], block adjustment of synchronizing signal (BASS) [7], and block processing [5].

The code- and carrier-tracking performance can be measured by the noise variance and the mean time to lose lock (MTLL). Increasing the PIT decreases the noise variance, which consequently increases the MTLL. If the bit edge position is identified and the data bit values are unknown, then the maximum PIT that can be used is the duration of one data bit interval (20 ms). The coherent integration time can be increased by either external or internal aiding. In external aiding, the data bit values are provided by an assisting source, such as wireless network. In internal aiding, the navigation message is decoded, and the repeated data are used to increase the coherent integration. A thorough discussion of the sources of tracking errors and the tracking threshold can be found in [9, Ch. 5].

In nonstationary dynamic receivers, the tracking technique must consider the receiver dynamics to achieve acceptable performance. The order of the tracking loop determines its ability to track different dynamics. The Doppler shift and Doppler rate are proportional to, respectively, the relative velocity and the relative acceleration between the GNSS satellite and the receiver. If one loop is used to track the carrier parameters (e.g., phase, Doppler shift, and Doppler rate), then a second-order loop can track velocity, a third-order loop can track acceleration, and a fourth-order loop can track jerk. Since GNSS satellites are continuously moving, at least a second-order tracking loop is required. If the receiver is accelerating, then a higher-order tracking loop will be required. In EKF approaches, the dynamic model determines the ability of the EKF to track different dynamics.

Tracking of high dynamics requires wider loop bandwidth. However, wider bandwidth means smaller PIT. Some GNSS receivers use velocity and acceleration aiding data. In this case, the dynamics can be either subtracted from the received signal or provided to the tracking loop. If the dynamics are subtracted from the received signal, then smaller loop bandwidth can be used, which decreases the noise variance and increases the MTLL. The dynamic aiding data can be provided by means like an inertial navigation system (INS). If the receiver dynamics are predictable, then a dynamic model can be used to obtain the velocity and acceleration data. An example of this is a GNSS receiver on a satellite where Kepler equations can be used to calculate the receiver's dynamics.

As discussed in [9, Ch. 5], the dominant sources of range errors in a DLL are the thermal noise range jitter and dynamic stress error. The DLL thermal noise code tracking jitter is

$$\sigma_\tau = \sqrt{\frac{8F_1\,\Delta_{EL}^2\,B_n}{S/N_0}\left(1-\Delta_{EL}+\frac{2F_2\,\Delta_{EL}}{T_I\,S/N_0}\right)} \text{ chips} \tag{6.3}$$

or

$$\sigma_\tau = \lambda_c\sqrt{\frac{8F_1\,\Delta_{EL}^2\,B_n}{S/N_0}\left(1-\Delta_{EL}+\frac{2F_2\Delta_{EL}}{T_I\,S/N_0}\right)} \text{ m} \tag{6.4}$$

where F_1 is a DLL discriminator correlator factor; Δ_{EL} is the separation between the early and late signals; B_n is the noise bandwidth of the loop; T_I is the PIT;

F_2 is a DLL discriminator type factor; and λ_c is equal to 293.05 m/chip for the GPS C/A code. The dynamic stress error is

$$R_e = \frac{dR^n/dt^n}{\omega_0^n} \text{ chips} \tag{6.5}$$

where ω_0 is the loop natural frequency, and dR^n/dt^n is expressed in chips/s^n, assuming dedicated early and late correlators ($F_1 = 1/2$), an early late-type discriminator ($F_2 = 1$), a second-order loop with $B_n = 10$ Hz ($\omega_0 = 10/0.53 = 18.87$), and 1g acceleration ($dR^2/dt^2$ = (9.8 m/s^2)/(293.05 m/chip) = 0.033 chip/s^2). Thus, (6.4) becomes

$$\sigma_\tau = 293.05\sqrt{\frac{40\,\Delta_{EL}^2}{S/N_0}\left(1 - \Delta_{EL} + \frac{2\Delta_{EL}}{T_I\, S/N_0}\right)} \text{ m} \tag{6.6}$$

and $R_e = 0.033/18.87^2 = 0.0001$ chip.

The performance of the code-tracking algorithm developed in Chapter 5, Section 5.3, is compared to the performance of a conventional tracking loop. Two cases are considered for the conventional tracking loop: (1) assuming the bit edge position is known, the PIT is 20 ms, and (2) assuming the data bit values are provided by an aiding source, the PIT is 400 ms. In each case, two code-delay separations are used: 1/2 and 1/5 chip. This is compared against the EKF that uses two code-delay separations of about 1/2 and 1/5 chip, as described in Chapter 5, Section 5.6.2. Figure 6.8 shows the first case in which the PIT is 20 ms

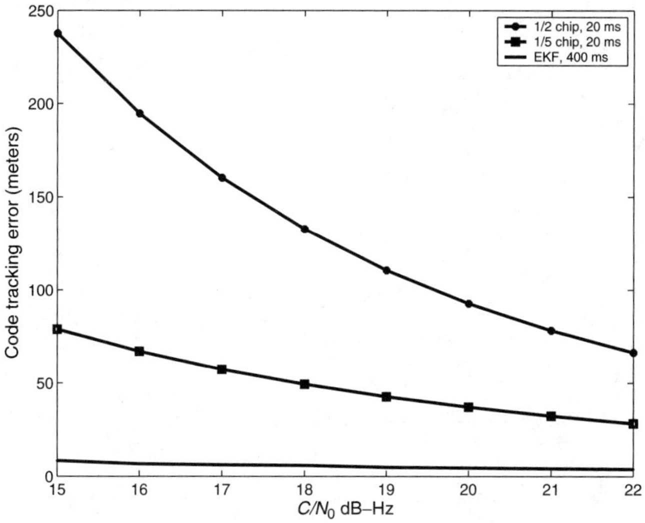

Figure 6.8 Code-tracking performance comparison between a conventional tracking loop with a PIT of 20 ms and the EKF code tracking presented in Chapter 5, Section 5.3.

for the conventional tracking loop. Figure 6.9 shows the second case in which aiding data are available, and the PIT is 400 ms for the conventional tracking loop. As can be seen from the figure, the developed EKF algorithm gives apparent better tracking performance.

As discussed in [9, Ch. 5], the dominant sources of frequency errors in an FLL are the thermal noise frequency jitter and dynamic stress error. The FLL thermal noise frequency tracking jitter is

$$\sigma_f = \frac{1}{2\pi T_I}\sqrt{\frac{8B_n}{S/N_0}\left(1+\frac{1}{T_I\, S/N_0}\right)}\ \text{Hz} \tag{6.7}$$

The 3-sigma dynamic stress error is

$$R_e = \frac{1}{360\,\omega_0^n}\frac{dR^{n+1}/dt^{n+1}}{\omega_0^n}\ \text{Hz} \tag{6.8}$$

assuming a first-order tracking loop with $B_n = 10$ Hz ($\omega_0 = 10/0.53 = 18.87$ Hz) and 1g maximum acceleration $(dR^2/dt^2 = 9.8 \times 360/0.1903 = 18,539°/s^2$.

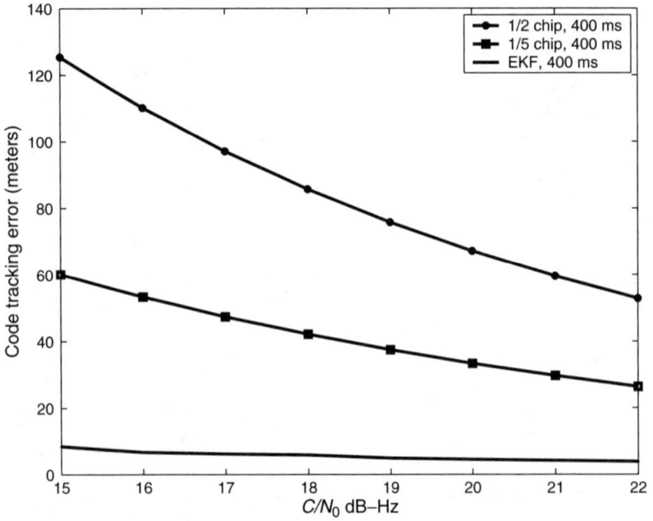

Figure 6.9 Code-tracking performance comparison between a conventional tracking loop with a PIT of 400 ms and the EKF code tracking presented in Chapter 5, Section 5.3.

Thus, $R_e = 2.7292$ Hz, and the 1 sigma dynamic stress error is 0.91 Hz. The performance of the EKF frequency tracking, using TCXO and OXO clocks, presented in Chapter 5, Section 5.4, is compared to the performance of a conventional tracking loop. Assuming the bit edge position is known, the PIT is 20 ms. Figure 6.10 shows the comparison result. As can be seen from the figure, the developed EKF algorithm gives better frequency-tracking accuracy.

As discussed in [9, Ch. 5], the dominant sources of phase errors in a PLL are the thermal noise phase jitter and dynamic stress error. The PLL thermal noise phase tracking jitter is

$$\sigma_\theta = \frac{360}{2\pi}\sqrt{\frac{B_n}{S/N_0}\left(1 + \frac{1}{2\,T_I\,S/N_0}\right)}\ ^\circ \tag{6.9}$$

Assume the dynamic stress error is negligible. The performance of the EKF phase tracking, using TCXO and OXO clocks, presented in Chapter 5, Section 5.4, is compared to the performance of a conventional tracking loop. Assume the bit edge position is known and the PIT is 20 ms. Figure 6.11 shows the comparison result. As can be seen from the figure, the developed EKF algorithm gives better phase-tracking accuracy.

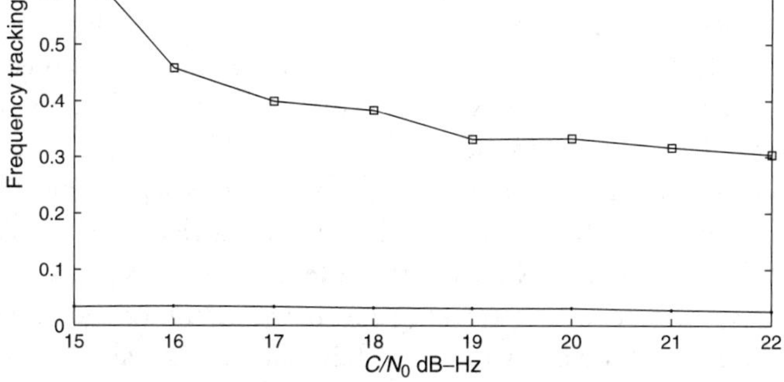

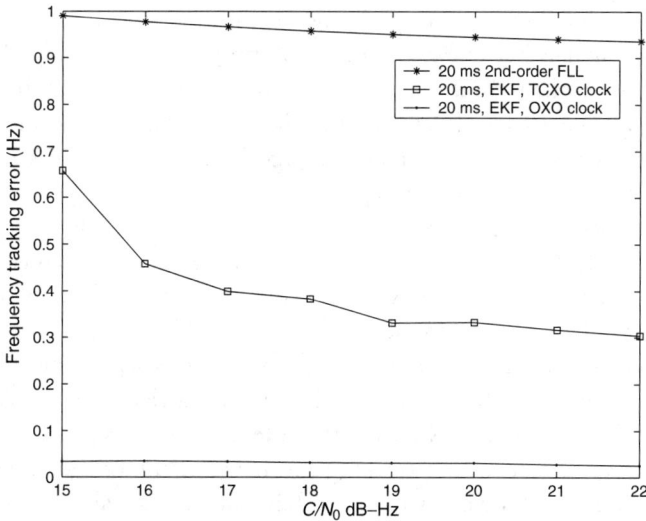

Figure 6.10 Frequency-tracking performance comparison between a conventional tracking loop with a PIT of 20 ms and the EKF frequency tracking presented in Chapter 5, Section 5.4.

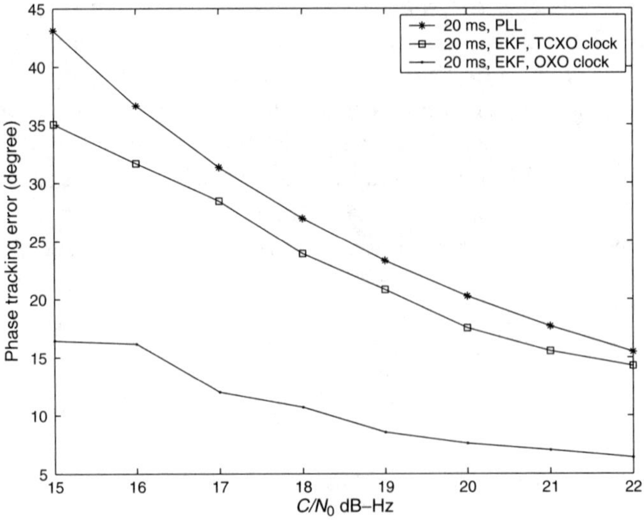

Figure 6.11 Phase-tracking performance comparison between a conventional tracking loop with a PIT of 20 ms and the EKF phase tracking presented in Chapter 5, Section 5.4.

6.5 New Civil L2C and L5 GPS Signals

The GPS is introducing new civil signals on the L2C and L5 carrier frequencies. The new signals have new structures and properties, including longer PRN codes, faster transmission rates, better cross-correlation properties, better multipath rejection, dataless signals, convolutionally encoded navigation messages, and navigation messages with new format. The algorithms developed in this book can be modified to work with the new signals.

In [10], acquisition and fine acquisition algorithms are developed to work with weak L2C and L5 signals. The acquisition of the L2C signal consists of three stages: (1) acquisition of the (CM) signal; (2) fine acquisition of the phase, Doppler shift, and Doppler rate; and (3) acquisition of the (CL) signal. The acquisition of the I5 and Q5 channels of the L5 signal is performed separately. The dataless Q5 channel is used for the fine acquisition of the carrier. New versions of the MDBZP are introduced to utilize the structures of the new signals.

In [11], code- and carrier-tracking and navigation message decoding algorithms are developed to work with weak L2C and L5 signals. The L2C tracking combines the CM and CL signals to form the integration and the EKF equations. The L5 tracking can work in three different modes: (1) tracking the I5 channel only, (2) tracking the Q5 channel only, and (3) tracking the I5 and Q5 channels together. In addition, a method is developed to decode the convolutionally encoded navigation message. Once the navigation message is decoded, it is reencoded and passed to the tracking modules to be used as an internal aid to remove the effect of the data signs before the coherent integration is calculated.

References

[1] Psiaki, M. L., "Block Acquisition of Weak GPS Signals in a Software Receiver," *Proc. ION GPS*, Salt Lake City, UT, September 11–14, 2001, pp. 2838–2850.

[2] Parkinson, B., and J. Spilker, *Global Positioning System: Theory and Applications*, Washington, D.C.: AIAA, 1996.

[3] Psiaki, M. L., and H. Jung, "Extended Kalman Filter Methods for Tracking Weak GPS Signals," *Proc. ION GPS*, Portland, OR, September 24–27, 2002, pp. 2538–2553.

[4] Psiaki, M. L., "Smoother-Based GPS Signal Tracking in a Software Receiver," *Proc. ION GPS*, Salt Lake City, UT, September 11–14, 2001, pp. 2900–2913.

[5] Lin, D. M., and J. B. Y. Tsui, "A Weak Signal Tracking Technique for a Stand-Alone Software GPS Receiver," *Proc. ION GPS*, Portland, OR, September 24–27, 2002, pp. 2534–2538.

[6] Gustafson, D. E., "GPS Signal Tracking Using Maximum-Likelihood Parameter Estimation," *Journal of Navigation*, Vol. 45, No. 4, 1999.

[7] Tsui, J. B. Y., M. H. Stockmaster, and D. M. Akos, "Block Adjustment of Synchronizing Signal (BASS) for Global Positioning System (GPS) Receiver Signal Processing," *Proc. ION GPS*, 1997, Kansas City, MO, Sept. 16–19, 1997, pp. 637–643.

[8] Zhodzishaky, M., et al., "Co-Op Tracking for Carrier Phase," *Proc. ION GPS*, 1998, Nashville, TN, Sept. 15–18, 1998, pp. 653–664.

[9] Kaplan, E., *Understanding GPS: Principles and Applications*, Norwood, MA: Artech House, 1996.

[10] Ziedan, N. I., "Acquisition and Fine Acquisition of Weak GPS L2C and L5 Signals under High Dynamic Conditions for Limited-Resource Applications," *Proc. ION GNSS 2005*, Long Beach, CA, September 13–16, 2005, pp. 1577–1588.

[11] Ziedan, N. I., "Extended Kalman Filter Tracking and Navigation Message Decoding of Weak GPS L2C and L5 Signals," *Proc. ION GNSS 2005*, Long Beach, CA, September 13–16, 2005, pp. 178–189.

Acronyms

AGC	Automatic gain control
AGPS	Assisted Global Positioning System
AWGS	Additive white Gaussian noise
BASS	Block adjustment of synchronizing signal
BER	Bit error rate
BOC	Binary offset carrier
BPSK	Binary phase shift keying
C/A	Coarse/acquisition
CASM	Coherent adaptive subcarrier modulation
CCMDB	Circular correlation with multiple data bits
CDMA	Code Division Multiple Access
CFI	Correct frequency initialization
CFT	Correct frequency initialization
C/N_0	Carrier-to-noise ratio
CS	Commerical service
CTR	Convergence and tracking rate
DA	Data aided
DBZP	Double block zero padding
DD	Data directed
DLL	Delay lock loop
DOP	Dilution of precision
DS/SS	Direct sequence spread spectrum
DTTL	Data transition tracking loop
EDR	Edge detection rate
EKF	Extended Kalman filter
ESA	European Space Agency
ES-VA	Extended states Viterbi algorithm

EU	European Union
FCR	Frequency convergence rate
FEC	Forward error correction
FEER	Frequency estimate with estimated rate
FEZR	Frequency estimate with zero rate
FFT	Fast Fourier transform
FLL	Frequency lock loop
FRE	Frequency and rate estimate
GLONASS	Global Orbiting Navigation Satellite System
GNSS	Global Navigation Satellite System
GPS	Global Positioning System
GS	Gaussian second order
HEO	High-Earth orbit
HOW	Handover word
IF	Intermediate frequency
IFFT	Inverse fast Fourier transform
INS	Inertial navigation system
ION	Institute of Navigation
KF	Kalman filter
LNA	Low-noise amplifier
MAP	Maximum a posteriovi
MDBZP	Modified double block zero padding
MEO	Medium-Earth orbit
ML	Maximum likelihood
MS	Mobile station
MTLL	Mean time to lose lock
NAV	Navigation message
NDA	Non–data aided
OS	Open service
OXO	Ovenized crystal oscillator
PDF	Probability density function
PIT	Predetection integration time
PLL	Phase lock loop
PRN	Pseudorandom noise
PRS	Public regulated service
QPSK	Quadrature phase shift keying
RF	Radio frequency
SAR	Search and rescue
SNR	Signal-to-noise ratio
SOL	Safety of life
TCXO	Temperature-compensated crystal oscillator
TLM	Telemetry

TOW	Time of week
TS	Truncated second order
TTFF	Time to first fix
VA	Viterbi algorithm
VCO	Voltage control oscillator
UHF	Ultra High Frequency
WGN	White Gaussian noise

Symbols

A	Received signal level
$\hat{A}$	Estimated received signal level
$\bar{A}$	Expected received signal level
C	Received C/A code
C_L	Local replica C/A code
d	Data bit value
E_{change}	Amount of change in the current E_{fz}
E_{fz}	Fraction of the current Doppler bins to be eliminated
E_{limit}	Maximum or minimum allowed value of E_{fz}
f_{code}	C/A frequency
f_d	Doppler shift
$\hat{f}_d$	Estimated Doppler shift
f_{dc}	Doppler shift on the C/A code
f_e	Doppler shift estimation error
f_{IF}	IF carrier frequency
f_{L1}	L1 carrier frequency
f_s	Sampling rate
H	EKF linearization matrix
I	In-phase correlator output signal
I_L	In-phase local signal
K	EKF gain
N_τ	Number of code-delay bins
N_b	Number of possible bit edge positions
N_{change}	Acquisition step number after which the current N_{ez} is to change
$N_{d_{min}}$	Minimum allowed number of Doppler bins in the acquisition
N_e	Number of steps between Doppler bin eliminations in the acquisition

N_{elim}	Number of times the value of N_{ez} changes
N_{ez}	Number of acquisition steps between two elimination steps
$N_{ez_{change}}$	Amount of change in N_{ez}
N_{fd}	Number of Doppler bins
N_{freq}	Number of possible frequencies used in the ES-VA
n_I	In-phase correlator output noise
$N_{I_{cMax}}$	Maximum number of data bits to be used in the carrier-tracking integration
$N_{I_{cMin}}$	Minimum number of data bits to be used in the carrier-tracking integration
$N_{I_{Max}}$	Maximum number of data bits to be used in the code-tracking coherent integration
N_{incoh_v}	Number of incoherent accumulations in the vth code-tracking step
n_Q	Quadrature-phase correlator output noise
N_{range}	Number of small Doppler ranges in DBZP
N_t	Number of data bit intervals in the coherent integration of the acquisition
P	EKF covariance matrix
p_d	Probability of detection
p_{DC}	Probability of detecting an interfering signal
p_f	Probability of false alarm
Q	Quadrature-phase correlator output signal
Q_L	Quadrature-phase local signal
Q_n	EKF process noise matrix
$\hat{R}(.)$	Autocorrelation function
R_n	EKF measurement noise matrix
Res	EKF residual
S	EKF square root covariance matrix
T_c	C/A code length (1 ms)
T_{car_u}	Integration length of the uth carrier-tracking step
T_{c_d}	C/A code length modified by the Doppler effect
T_{chip}	Code chip length
T_{c_i}	Code length modified by the Doppler effect at the ith interval
T_{code_v}	Integration length of the vth code-tracking step
T_d	One data bit length
T_{d_m}	Data length modified by the Doppler effect at the mth interval
t_{first_m}	Time of the first sample in the mth interval
T_I	Predetection integration time
t_k	Sampling time
T_s	Time between samples
t_{s_m}	Difference between t_{first_m} and t_{start_m}
t_{start_m}	Time of the start of the first PRN code in the mth interval

T_t	Total integration time for acquisition
$\boldsymbol{x}$	EKF states
W_{f_d}	Clock frequency disturbance
W_θ	Clock phase disturbance
$\boldsymbol{z}$	EKF measurement vector
$\hat{\boldsymbol{z}}$	EKF expected measurement vector
α	Doppler rate
$\hat{\alpha}$	Estimated Doppler rate
α_e	Doppler rate estimation error
δ	Signed code-delay separation
Δ	Code-delay separation (prompt-late and prompt-early)
γ	Acquisition threshold
ϕ	Possible bit edge position
Φ	EKF transition matrix
τ	Code delay
$\hat{\tau}$	Estimated code delay
τ_e	Code-delay estimation error
θ	Phase
$\hat{\theta}$	Estimated phase
θ_e	Phase estimation error

About the Author

Nesreen I. Ziedan received a Ph.D. in electrical and computer engineering from Purdue University, United States, in December 2004; an M.S. in control and computer engineering from Mansoura University, Egypt, in March 2000; an IT diploma in computer networks from the Information Technology Institute, Egypt, in September 1998; a B.S. (with honors) in electrical engineering (communications and electronics engineering) from Zagazig University, Egypt, in July 1997, where she ranked first in the faculty of engineering.

Ziedan is a research scientist at the University of Calgary, Canada. She is also a tenured assistant professor at the Computer Systems and Engineering Department, Faculty of Engineering, Zagazig University, Egypt. Her previous experience includes work as a research associate at Miami University, United States, in 2005, a consultant in 2004–2005, and a lecturer at the Computer Systems and Engineering Department, Faculty of Engineering, Zagazig University, Egypt, from January 1998 to December 1999.

Ziedan's biography is published in the 2004–2005 edition of the Chancellor's List, United States. She was awarded a governmental fellowship that supported her entire Ph.D. at Purdue University. She was awarded a scholarship from the Information Technology Institute, Egypt, in 1997–1998. She received a number of awards and certificates of appreciation. She was awarded the Distinguished Egyptian Association Award and an honorary membership in 1997.

Ziedan has worked in many research areas, including Global Navigation Satellite System (GNSS) receiver design, deep integration of GNSS with inertial navigation systems (INS), the design and implementation of network cards and software for local area networks, asynchronous transfer network (ATM) control, congestion control and management, the application of artificial intelligence and neural networks to communications and computer problems, and computer security and denial of service.

Ziedan has several (pending) patents from her work in the design of GNSS receiver algorithms for weak signals.

Index